AI for Beginners

Learn Artificial Intelligence Now and Stay Ahead of the Curve

Cyber Safety Net
3450 Palmer Drive #4-286
Cameron Park, CA 95682 USA
mark@cybersafetynet.net
(844) 580-1200

20260325

Hey there, cool cats! Mr. Sugartooth here, ready to take you on a wild ride into the world of AI (artificial intelligence), which Dictionary.com defines as "the capacity of a computer, robot, programmed device, or software application to perform operations and tasks analogous to learning and decision making in humans, such as speech recognition or question answering."[i] Buckle up! We're about to make learning AI as fun as driving fast with the wind blowing through your hair! First time? That's okay. Go slow for a bit then... you know.

AI is a Supercharged Research Tool

Say Hello to Your New Research Assistant

Search Engine on Steroids

Start by viewing AI, or artificial intelligence, as a research tool and writing aid. I prefer Perplexity.ai. It is like having a super-smart mentor who knows everything about everything. It's an artificial intelligence tool that can answer your questions, help with proposals, or just chat about the future. Think of it as a search engine on steroids. I do.

Perplexity.ai is not the only AI service out there. It is the one I'll be covering here. I like it because it cites the sources of its research. This is like someone telling you to drink 15 cups of water every day and you respond, "Woah! Who says I should do *that*?"

Where You Find It

First things first, head over to https://perplexity.ai. You'll find a search bar right in the middle of the page. That's where the magic happens! Just type in your question or topic, hit enter and watch Perplexity work its magic.[ii]

The Power of Follow-ups

Follow-ups Lead to More Valuable Answers

Now, here's where AI gets really user friendly. Perplexity.ai has a valuable feature called follow-ups. You ask a question, get an answer and then dig deeper. Let me break it down for you:

- Ask your first question
- Read the answer Perplexity gives you
- Click the "Ask follow-up" button
- Ask another question related to the first one
- Keep going until you're satisfied with the info.

I am using Perplexity Pro, available at https://perplexity.ai. This costs $200 for an annual subscription. The free version is also valuable. Start there. I did. Let's see how prompting and follow-ups work.

How Much Water?

Here's my first prompting example: "How much water should a 50-year-old man drink each day?" Perplexity answers with:

For a 50-year-old man, the general recommendation for daily water intake is about 15.5 cups, which is equivalent to approximately 3.7 liters or 125 ounces of fluids per day. This guideline includes all fluids consumed, not just plain water. About 20% of this fluid intake typically comes from food, with the remaining 80% from beverages. However, individual water needs can vary based on several factors:

- Activity Level: Increased physical activity, especially in hot or humid conditions, can increase water needs due to higher sweat losses.
- Climate and Environment: Living in hot, dry, or high-altitude areas can increase fluid requirements.
- Health Conditions: Certain health issues or medications may affect hydration needs. conditions like diabetes or medications such as diuretics can increase fluid loss.
- Age: As people age, their sense of thirst can diminish, making it important to consciously ensure adequate fluid intake.

It's important to note that while these are general guidelines, the best approach is to listen to your body. Drinking enough water to maintain clear, pale urine and adjusting intake based on thirst and activity level are practical strategies....

It is HOT Outside

Follow-up this answer. You can add more info to create a more valuable answer. In the same thread, the prompt box presents, "Ask follow-up." OK. Let's do that. Just type: "**and** the man works outside **and** the high temperature this week will exceed 100 degrees Fahrenheit." Perplexity answers with:

When working outside in high temperatures, such as those exceeding 100 degrees Fahrenheit, it is crucial to increase water intake to prevent dehydration and heat-related illnesses. For a 50-year-old man working in such conditions, the following hydration guidelines are recommended:

- Drink Regularly: Consume approximately 1 cup (8 ounces) of water every 15 to 20 minutes. This equates to about 0.75 to 1 quart (24 to 32 ounces) per hour.
- Avoid Overhydration: Do not exceed 48 ounces (1.5 quarts) of water per hour to avoid hyponatremia, a condition caused by low sodium levels in the blood due to excessive water intake.
- Electrolyte Replacement: Consider drinking beverages with electrolytes, such as sports drinks, especially during prolonged periods of sweating, to replenish lost minerals like sodium and potassium.

Additionally, it is important to take regular breaks in shaded or cooler areas and to monitor for signs of dehydration and heat-related illnesses, such as dizziness, fatigue, or muscle cramps. Adjusting work schedules to cooler parts of the day and wearing appropriate clothing can also help mitigate heat stress.

Go one step further. The previous answer mentioned "appropriate clothing." What exactly does "appropriate clothing" mean? Follow-up by asking Perplexity: "**and** what recommendations do you have for appropriate clothing?" Perplexity answers with:

Here are some key recommendations for appropriate clothing:

Fabric Choices

- Opt for lightweight, breathable fabrics like cotton, linen, or moisture-wicking synthetic blends
- Avoid heavy or tight-fitting clothing that can trap heat
- Choose fabrics with high air permeability to allow better airflow and heat dissipation

Clothing Features

- Select loose-fitting garments to promote air circulation
- Wear light-colored clothing to reflect sunlight and heat
- Look for clothes with moisture-wicking properties to help evaporate sweat quickly

Specific Clothing Items

- Lightweight, breathable long-sleeved shirts for sun protection
- Wide-brimmed hats or hats with "sun-skirts" to shade the face, neck and ears
- Breathable, moisture-wicking socks
- If possible, consider clothing with built-in UV protection

Additional Protective Gear

- Use sunglasses with 100% UV protection
- Apply water-resistant sunscreen with at least SPF 30

Workplace-Specific Considerations

- If high-visibility clothing is required, choose breathable options designed for hot weather
- For jobs requiring protective equipment, look for vented options when possible

Remember to stay hydrated and take regular breaks in shaded or air-conditioned areas when working in extreme heat. Be aware of the signs of heat-related illness and adjust work schedules if possible to avoid the hottest parts of the day.

You probably weren't expecting all this valuable information with your initial "How much water should a 50-year-old man drink each day?" prompt. That's the power of the follow-up. Your initial prompt can be general. You then follow-up with more specific prompts to reach more valuable answers.

In case you're wondering...it is 108 degrees in Sacramento as I write this. I am drinking a lot of water today. Chug. Chug. Chug.

Reading Level

You can write (or let your AI engine write) content and check the Flesh-Kincaid Grade Level. Wikipedia has a fantastic article on Flesh-Kincaid.[iii] Basically, the shorter the average word and average sentence, the easier the text is to read. The opposite is also true.

Getting back to business... type: "**and** respond at a 6th grade reading level." Perplexity answers with:

When it's really hot outside, like over 100 degrees, and you're working outside, it's super important to drink more water to stay safe and healthy. Here are some simple tips:

- Drink Often: Try to drink about a cup of water every 15 to 20 minutes. That's about 3 to 4 cups every hour.
- Don't Drink Too Much: It's important not to drink more than 6 cups of water in an hour so you don't get sick from too much water.
- Electrolytes: Sometimes, it's good to drink sports drinks. They help replace important things like salt that you lose when you sweat a lot.

Also, make sure to take breaks in the shade and watch out for signs that you might need more water, like feeling dizzy or tired. If you can, try to work during the cooler parts of the day and wear light, breathable clothes.

Just easier and more friendly, right? That's because this verbiage is for those reading at a 6th grade reading level. If you want your AI engine to explain things for 6th graders, just ask! Want it to explain things for 12th graders, just ask. Type: "**and** respond at a 12th grade reading level." Fasten your seatbelt. Perplexity answers with:

In extreme heat conditions exceeding 100°F, proper hydration becomes crucial for individuals engaged in outdoor labor to maintain safety and optimal health. Consider the following comprehensive guidelines for effective hydration:

Hydration Strategy

Frequency and Volume: Aim to consume approximately 8 ounces (237 ml) of water at 15 to 20-minute intervals. This translates to a total fluid intake of 24 to 32 ounces (710 to 946 ml) per hour.

Moderation: While staying hydrated is essential, it's equally important to avoid excessive fluid consumption. Limit your hourly water intake to no more than 48 ounces (1.4 liters) to prevent hyponatremia, a potentially dangerous condition caused by dilution of blood sodium levels.

Electrolyte Balance: In situations of prolonged sweating, consider supplementing your fluid intake with electrolyte-rich beverages. These drinks can help replenish crucial minerals, particularly sodium, that are lost through perspiration.

Additional Precautions

To further mitigate the risks associated with working in high-temperature environments:

- Incorporate regular rest periods in shaded areas to allow your body to cool down and recover.
- Remain vigilant for signs of heat-related illness, such as dizziness, fatigue, or disorientation.
- When possible, schedule strenuous activities during cooler parts of the day to minimize heat exposure.
- Opt for lightweight, breathable attire that facilitates evaporative cooling and protects against direct sunlight.

By adhering to these guidelines, you can significantly reduce the risk of heat-related illnesses and maintain your productivity in challenging thermal conditions.

Just easier and more friendly, right? Haha, just kidding. More difficult. This one is at a 12th grade reading level.

What do you choose? A report at the 6th grade reading level may appear simplistic. A report at the 12th grade reading level may go over the heads of your readers. Your decision.

You may never stop using follow-ups. However, I think you will find that as you get more comfortable with AI, your initial prompts become more thorough. You can combine prompts to achieve a quicker and more thorough answer. The "How much water" cumulative prompt would read, "How much water should a 50-year-old man drink each day **and** the man works outside **and** the high temperature this week will exceed 100 degrees Fahrenheit **and** what recommendations do you have for appropriate clothing **and** respond at a 6th grade reading level."

Spelling counts. However, AI forgives repeated usage of "and," incorrect punctuation and run-on sentences. That's not what we learned in school. If my high school English teachers ever read the cumulative prompt (above) on a paper I submitted, they would have pushed me off the roof while yelling, "We thought he was paying attention!"

Like I'm 10

Flesh-Kincaid Grade is rather intellectual. Another strategy is to specify an age in your prompt. Your prompt could end with "**and** respond like I am 10 years old." Remember...you're learning. It is ok to get overly simplified answers when you begin your research.

And And And

English Teachers Say NO but AI Says YES

Perplexity's tolerance of poor grammar makes life easier, wouldn't you say? Here's another example of and and and run on sentences as we learn about the CIA Triad in cybersecurity.

The CIA Triad

Let's say you're curious about the CIA Triad in cybersecurity. No, it's not about spies, but it's still valuable. Start by asking Perplexity about the CIA. Then, use follow-ups to make the answer even better. Follow this prompting sequence in Perplexity.

- "Tell me about the CIA" and press (Enter). What do you get?

- “Tell me about the CIA Triad in cybersecurity” and press (Enter). What do you get?
- "**and** write at a 6th grade reading level" and press (Enter). What do you get?
- "**and** in under 500 words" and press (Enter). What do you get?
- "**and** give me this information in bullet points" and press (Enter). What do you get?
- "**and** make sure each bullet point list has only three entries" and press (Enter). What do you get?
- "**and** bold the first word in each bullet point entry" and press (Enter). What do you get?

See how each "and" adds something new? Your initial prompt does **not** have to be fantastic.

I also told Perplexity.ai how many words I wanted. You can specify a minimum, maximum or range of words. A sample prompt would read, “Tell me about the CIA Triad in cybersecurity **and** in more than 250 words **and** less than 500 words.”

Think Twice When Prompting

Now, as your cool teacher, I got to remind you to stay safe online. AI is awesome, but the Internet has some scary places. Type your prompt and think twice before pressing (Enter). Here are some things to keep in mind:

- Don't share personal info. This includes your home address, social security number, school ID number, embarrassing pictures and health information. Items you add in prompts today could end up in somebody else’s answers tomorrow. AI is smart, because it is always learning. If you submit personal information in your prompts, it will go into the pool of what AI knows. What you thought was private may no longer be private.
- Double-check important facts with other sources. It is ok to use multiple AI services.

- Use your critical thinking skills. AI is smart, but it's not perfect! I did a search for the Lotus company. I was thinking about the company that published a product called Lotus 123. When I typed my prompt and hit (Enter), I received information about Lotus Cars Limited. I had to refine and narrow my prompt. You know where this is going...I had to **follow-up** my initial prompt to learn what happened to the company that published Lotus 123.

Homework

Time for some fun **and** to put this knowledge to use. Like Mark says in his cybersecurity classes, "Life is not a spectator sport."

Nighttime Baseball

Nighttime baseball is no big deal these days. It was not always that way at Wrigley Field. Here is your homework, sports fans:

"Write 250-500 words about Wrigley Field. Tell me where it is, who plays home games there, when the first night game occurred, which two teams played, what the final score was and why Wrigley Field was the last MLB stadium to allow night games." Send your response to mark@cybersafetynet.net.

Golden Gate Bridge

The Golden Gate Bridge may be one of the world's most photographed pieces of art. Having ridden my bike across the Golden Gate Bridge, I would say the pictures do not do it justice. Regardless, I am adding a picture here. You should visit and experience the Golden Gate Bridge for yourself. Here is your homework, architecture fans:

"Write 500-1,000 words on the Golden Gate Bridge. Tell me where it is, which two counties it connects, what color it is, who built it, when it was built, how tall it is, how long it is, how many cars travel on it every day and how I would get to it from Salt Lake City, by car." Send your response to mark@cybersafetynet.net.

Halloween

When can you dress up like Dracula and go to school or work? Only during Halloween season. Time to pull those Halloween decorations and costumes down from the attic. Different people celebrate in their own ways. What can you tell me about Halloween? Here's your homework, Dracula fans:

"Write at least 500 words on Halloween. Tell me how different countries celebrate Halloween, why we celebrate it on October 31, why we wear costumes, why we knock on neighbors' doors and yell **TRICK OR TREAT** and why pumpkins are so popular." Send your response to mark@cybersafetynet.net.

Extra Credit

I have not mentioned anything about reading level in these homework assignments. You can add follow-ups that request answers at a specific reading level.

Section Summary

Using tools like Perplexity.ai is like stepping into the future. You're not just searching for information; you're having a conversation with artificial intelligence. The powerful thing about Perplexity is that you can make it work just the way you want. Need a shorter answer? Ask for it! Need a longer answer? Ask for it!

Just like learning how to throw a baseball, you get better with practice. Keep these tips in mind as you gain experience and comfort with AI.

- Be specific: The more details you give, the better answers you'll get
- Use follow-ups: Don't be shy about asking for more info or clarification
- Experiment with different phrasings: Sometimes, asking the same question in a different way can give you new insights
- The "and" technique helps Perplexity understand that you want to build on previous answers
- Ask for different formats, like bullet points, charts or tables

Section Quiz

Life is not a spectator sport. Here's a quiz.

1. **What's the main advantage of using follow-ups in Perplexity.ai?**

a) It makes the answers longer
b) It allows you to refine and customize your answers

c) It automatically puts headings in ALL CAPS
d) It corrects spelling mistakes

2. **How can using the word "and" in follow-up prompts help?**
a) It makes Perplexity respond faster
b) It tells Perplexity to forget previous information
c) It helps Perplexity build on previous answers
d) It automatically cites its sources

3. **What is a good practice when using Perplexity.ai?**
a) Always accept the first answer without questioning
b) Use follow-ups and be specific in your questions
c) Only use it for one subject
d) Use it for pictures

4. **What is Perplexity primarily described as in the text?**
a) A search engine
b) A super-smart friend who knows everything
c) A homework solver
d) A voice to text service

5. **What feature of Perplexity.ai does the text emphasize for its user-friendliness?**
a) Voice recognition
b) Image search
c) Music search and creation
d) Follow-up questions

6. **According to the text, what is the recommended approach when using Perplexity?**
a) Ask only one question at a time
b) Use complex language for better results
c) Start with general prompts and follow up with specifics
d) Write prompts that would make your high school English teachers proud

7. **What does the text suggest about Perplexity.ai's tolerance for grammar and punctuation in prompts?**
a) It requires perfect grammar and punctuation
b) It doesn't care about correct punctuation or run-on sentences
c) It only accepts prompts with proper sentence structure
d) Like your high school English teacher, you have to get everything right

8. **What safety tip does the text emphasize when using Perplexity?**
a) Always use a VPN
b) Don't share personal information
c) Limit usage to 30 minutes per day
d) Use voice to text to ensure confidentiality

9. **How does the text suggest users can control the reading level of Perplexity.ai's responses?**
a) By adjusting settings in the user profile
b) By requesting a specific reading level in the prompt
c) Use voice to text for improved accuracy
d) It's not possible to control the reading level

10. **Why should you do the homework?**
a) You have too much free time
b) You need an excuse to stare at the computer
c) You can show off what you know to your friends
d) Life is not a spectator sport

Answers

1. b) It allows you to refine and customize your answers
2. c) It helps Perplexity build on previous answers
3. b) Use follow-ups and be specific in your questions
4. b) A super-smart friend who knows everything
5. d) Follow-up questions

6. c) Start with general prompts and follow up with specifics
7. b) It doesn't care about correct punctuation or run-on sentences
8. b) Don't share personal information
9. b) By requesting a specific reading level in the prompt
10. d) Life is not a spectator sport

Do Better Research
Write Better Content

Better Prompts Create Better Answers

I hope you enjoyed the Golden Gate Bridge homework assignment. Let's talk some more about bridges. When you're prompting Perplexity (or any AI engine), be specific. Instead of prompting, "Tell me about bridges," try something like "What are the engineering challenges of building suspension bridges?" The more specific you are, the better answers you'll get. It's like ordering pizza. Tell them exactly what toppings you want, or you might end up with anchovies. Nuf said.

Perplexity has some fantastic features that'll make your research more fun. You can upload files, choose a specific focus for your questions and even attach extra info to your queries. Even more, Perplexity gives you citations for the info it provides. Other AI sources cannot make that claim.

Up-to-date Information

One of the best things about Perplexity is that it gives you up-to-date info. It's not stuck in the past like your dad's old encyclopedias. Whether you're researching current events or the latest tech, Perplexity has the freshest facts. Just remember, while Perplexity is super smart, it's not perfect. Always double-check what your computer tells you!

Choosing Your Topic

Overcoming Writer's Block

Let's talk about picking the perfect topic. You know how sometimes you stare at a blank page and your mind goes totally blank? Well, that was me this morning.

Getting back to business...Perplexity AI is here to rescue you from the dreaded writer's block! It is like having a brainstorming buddy who never runs out of ideas.

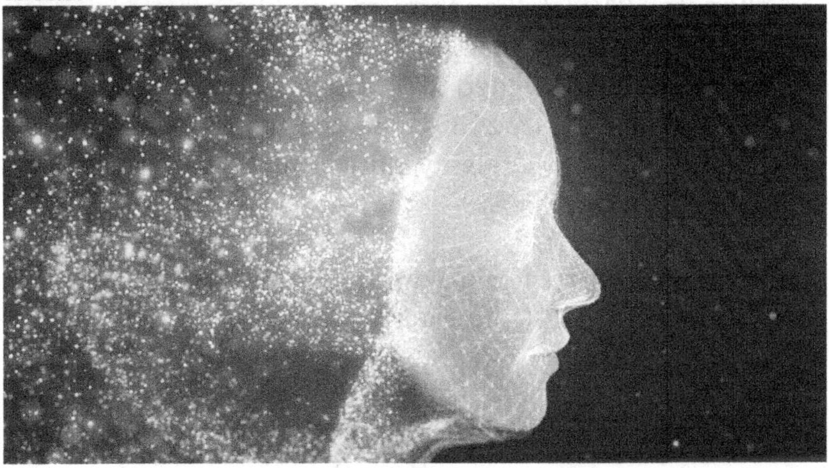

Use Perplexity to check out what's trending right now. Type in something like "What are the current trending topics in science?" or "What are people talking about in the world of travel?" Perplexity will hit you with a tsunami of cool ideas. It's like surfing the web, but way faster and cooler!

I found these interesting articles after clicking Perplexity's Discover link this morning:

- **World's First Fusion Power Plant**: "Commonwealth Fusion Systems (CFS) has announced plans to build the world's first grid-scale commercial fusion power plant, known as ARC, in Chesterfield County, Virginia."

- **Notre Dame's Global Donors**: "The restoration of Notre Dame Cathedral, following the catastrophic 2019 fire, has been fueled by nearly $1 billion in global donations, including substantial pledges from French billionaires, significant contributions from American donors totaling $62 million, and support from small donors worldwide, underscoring the cathedral's universal cultural and historical importance."
- **EU Funds Starlink Rival**: "The European Union has allocated €10.6 billion through a public-private partnership to develop IRIS², a 290-satellite constellation designed to provide secure, high-speed internet globally by 2030, enhance European technological sovereignty, and compete with Elon Musk's Starlink."

...and many more. In case you find yourself doing some mindless web surfing, go visit https://www.perplexity.ai/discover and find something interesting.

Ask Follow-up Lets You Dig Deeper

Spoiler alert: I am revisiting follow-ups and the Golden Gate Bridge. Follow-ups are too valuable to mention just once.

Once you've created some general ideas, use Perplexity to dig deeper. Let's say you're interested in travel because you're tired of staying home all the time. Ask Perplexity, "What is that orange bridge in San Francisco?" Boom! You get the name. Next, follow-up by prompting Perplexity, "Can I ride my bike on the Golden Gate Bridge?"

Before you buy plane tickets to San Francisco, make sure your topic is doable. Use Perplexity to check if there's enough info out there about your chosen subject. Ask it to give you an overview of available research or key experts in the field. If Perplexity comes back with tons of info, you're golden. If it cannot find much, choose another topic.

Last but not least, use Perplexity's focus feature to explore different angles of your topic. The message "Ask follow-up" is a blessing in disguise. It is like putting on different pairs of glasses to see your subject in new ways. While researching, the Golden Gate Bridge, you could focus on its history, its engineering, its impact on San Francisco, its color and its appearance in movies. Perplexity can help you explore each angle, letting you choose the one you prefer!

Analyze Text and PDFs

Now that you have chosen a vibrant topic with tons of information online, it's time to build a game plan. Think of Perplexity as your personal research assistant. Start by asking it broad questions about your topic, then gradually get more specific. It's like peeling an onion, but without the tears.

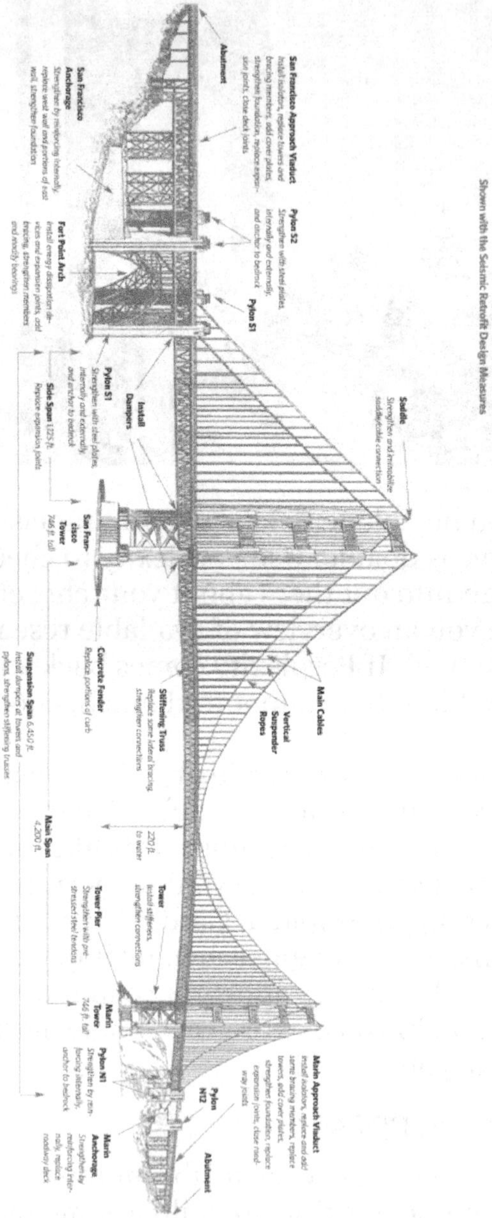

Perplexity lets you upload files for analysis. Got a PDF about your topic but don't want to read all 185 pages, or understand the whole thing? Feed it to Perplexity and ask it to give you the highlights. I did this with the Golden Gate Bridge retrofit drawing on the previous page.[iv] It's like having a study group inside your computer! Just remember, Perplexity is not reading your mind, so make sure to ask it **specific** questions about the document. Make your prompts overly deliberate. Follow-up often.

Even geniuses need to double-check their facts. That's where cross-referencing comes in. Use Perplexity to find multiple sources for your key info. If three different sources all say the Golden Gate Bridge is painted "International Orange," then you can bet good $$ it is.

Speaking of sources, Perplexity is great at helping you find credible content. When you're researching, look for info from universities, government agencies and well-known organizations. Perplexity can help you spot these reliable sources and avoid nonsense.

Lastly, don't forget that Perplexity is here to help you think, not to think for you. The goal isn't just to collect facts, but to understand them, learn and form your own ideas. Now that you have the facts, start writing your outline.

No Blind Faith

You cannot place blind faith in AI. Someone will find shortfalls in your AI-created work and cite your embarrassing faults. You **must** proofread.

I already referenced an AI prompt for the CIA. Although the example was the CIA Triad in cybersecurity, Perplexity (and most AI engines) will assume you are asking about the Central Intelligence Agency.

I used AI to generate content for a word search book. I began that project by asking AI to give me 100 words associated with Thanksgiving. The result was only 76 words. I kid you not. I followed up by asking for an additional 24 words.

In *Mastering AI-Powered Music*, I shared pictures of women playing piano and guitar. Some fantastic pictures simply did not make the cut. AI did not define four fingers on each hand. The cover art of *AI Driven Dynamic Pricing* showcases a woman who was the 7th candidate. Other artwork mangled the fingers. I am confident that someday, AI will figure out people have four fingers and one thumb on each hand.

AI Hallucinations

I was rather proud of my "no blind faith" label, then I read a LinkedIn article introducing the phrase *AI Hallucination*. This happens when AI generates a false answer and continues down a path assuming the false answer is valid. This can generate hilarious results.

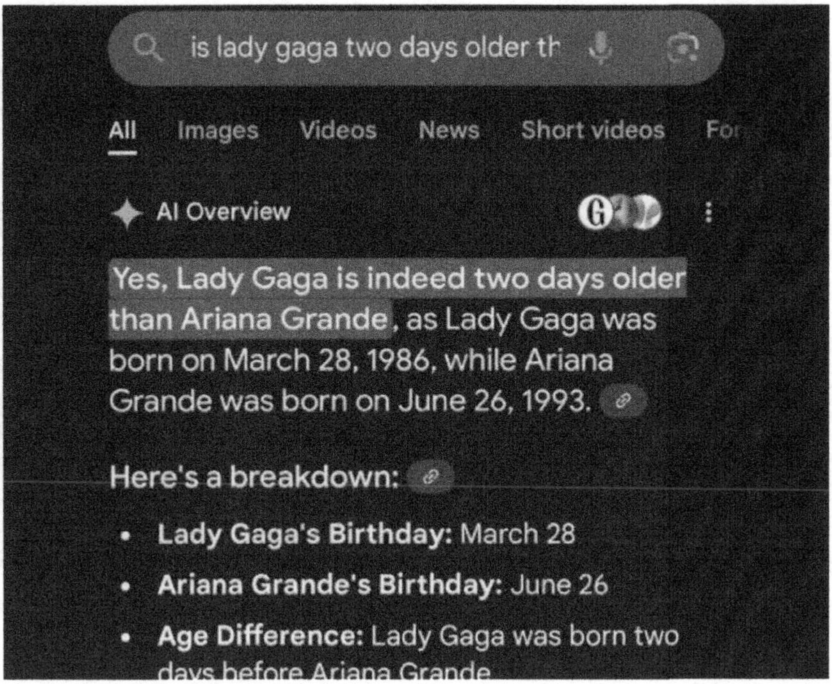

I have another AI hallucination in the next section. Whoops, I meant to write, "I will be *documenting* another AI hallucination in the DALL-E section." Just watch for a woman riding a bike on the Golden Gate Bridge. You'll know it when you see it.

Researching Aids Outlining Aids Writing

The Better the Outline, the Better the Result

Creating an outline is like drawing a map for your research adventure. And guess what? Our AI buddy Perplexity is going to be our GPS! First, ask Perplexity to give you the main points about your topic. For example, if you're writing about the Golden Gate Bridge, you might get topics like "history," "construction," and "cultural impact."

Take those main points and ask Perplexity to break them down further. It's like zooming in on a map to see the details. Under "history," you might find "why it was built" and "key people involved." Under "construction," you could find "engineering challenges" and "materials used."

Remember, though, this is your report. Feel free to shuffle topics. Maybe you want to start with the bridge's appearance in movies, then go back in time to its construction. Perplexity can help you flesh out each section, but you decide how it all fits together. It's like being the director of your own documentary.

International Orange Was Not in My Original Outline

You can also enhance your outline. While researching the Golden Gate Bridge, I learned about the color International Orange. Although that was not in the original outline, I am adding it to my outline. You can too.

Once you've got your main sections, it's time to add some pizzazz. Ask Perplexity for interesting facts or stories about each part. Maybe there's a cool tale about a daring worker who saved the day during construction, or a fun fact about how many paintbrushes it takes to keep the bridge looking fresh.

Write Your Report

Alright, wordsmiths, it's time to turn that outline into a full-blown report! Perplexity is going to be your writing partner, but remember, **you** are the director. Start by asking Perplexity to expand on each point in your outline. It's like having a super-smart friend explain things to you.

But here's the trick: don't just copy what Perplexity says. That's boring, and it won't sound like you. Instead, read what it gives you, then explain it in your own words. Imagine you're telling your best friend about what you learned. That's how your report should sound.

Don't be afraid to ask Perplexity for examples or fun facts to spice up your writing. Maybe there's a fun story about a dog that helped build the Golden Gate Bridge. Okay, I made that up, but people familiar with my writing would say it fits within my style. Make sure you polish AI answers to meet **your** style.

Now, let's talk about making your writing flow smoother than a chocolate milkshake. Use transition words to connect your ideas. Words like "however," "furthermore," and "consequently" can help your paragraphs flow into each other like a well-choreographed dance routine.

Don't forget to vary your sentence structure. Mix short, punchy sentences with longer, more detailed ones. It's like a roller coaster for your reader's brain. This keeps them engaged and excited to see what comes next.

Here's a pro tip: read your work out loud. If you stumble over a sentence, your readers will too. Smooth out those bumpy parts until your report flows smoothly.

Remember, writing involves rewriting. Don't expect your first draft to be perfect. That's why pencils have erasers and computers have Delete keys. Keep tweaking and polishing until your report shines brighter than a disco ball.

Write the Introduction Last

Yes, you read that right. Don't forget to plan your introduction and conclusion. Your intro should be like the trailer for a blockbuster movie. It should get people excited to read more. Your conclusion? That's your chance to wrap everything up with a bow and leave your readers thinking, "Wow, that was awesome!"

While I attended Michigan State University, a professor told me to write the body, the conclusion and the introduction, in that order. In list order, that would be:

1. The body
2. The conclusion
3. The introduction

I know this is a tall order. Perplexity can help you brainstorm ideas to make that happen.

Adding Visual Elements

Find Images That Support Your Writing

Picture your report, but with vibrant images that engage the reader. Adding images makes everything better. First up, use Perplexity to find out what kinds of images would work best for your topic. Then use Google Images. For the Golden Gate Bridge, you might want historical photos, diagrams of its structure, or even artistic renderings.

Some AI tools generate and host images. But hold your horses! You have to be careful with these. Always check if you're allowed to use AI-generated images for your report. If not, no worries! Perplexity can help you find real photos that are free to use. Just ask for "copyright-free images of the Golden Gate Bridge."

Create Images with ChatGPT and DALL-E

You can use ChatGPT and DALL-E to create images. I'm sure you will find this functionality helpful. However, sometimes the AI does not follow instructions. Here's why I say that.

In ChatGPT, I prompted:

"Give me a picture of a woman riding a bicycle on the Golden Gate bridge. There are no cars, people or other bicycles on the bridge. Have her riding toward us wearing an aqua helmet and dark blue cycling kit from Voler. She is riding a Giant TCR2 red and black road bike." ChatGPT answered with:

The first image from ChatGPT and DALL-E followed some instructions, but surely not all. When you're picking or creating images, think like a magazine editor. Choose pictures that grab attention and help explain your points. A photo of the bridge shrouded in fog could be perfect for talking about San Francisco's weather. A diagram of the bridge's cables could help explain its engineering. Make sure you double-check the pictures. AI picture generation is still an imperfect science. Make sure pictures have a strong enough resolution to remain sharp in case you increase their size. Yes. This qualifies as an AI hallucination.

Serious Researchers Cite Their Sources

Give credit where credit is due. Citing sources is like telling your readers, "Hey, I didn't just make this up. Check out where I learned it!" It shows you're a serious researcher, not just some random kid with a wild imagination.

Perplexity is a citation superstar. When it gives you info, it usually provides the source too. That's why you get little numbers in Perplexity's answers. Make sure you include that information in **your** report.

Why bother citing sources in the first place? Well, it's not just about avoiding plagiarism (which is theft, by the way). It's also about joining the conversation. You are showing how your ideas connect to what other smart people have said. Good citations can make your report stronger. They show you have done your homework and know what you're talking about. It's like having a bunch of experts backing you up. Your work gains credibility.

Disclosure Shows You Are Honest

Nobody knows everything. If you disclose what you know and what you don't, you gain more credibility with your reader. I could write a whole book on disclosure. Wait a sec, I did. See *Mastering AI Disclosure* on Amazon Kindle or continue to the *AI Disclosure* section (below).

Editing and Polishing

Let Perplexity Review Your Work

Okay, wordsmiths, you're in the home stretch! You have written your report. Use Perplexity to check for any sneaky grammar or spelling mistakes. It's like having a super-smart proofreader looking over your shoulder. Ask Perplexity for suggestions on how to make your writing more engaging. It can suggest snazzier words or help you vary your sentence structure. It gives your report a final polishing.

Let Your Work Sit Overnight, Then Review It

Here is an often-ignored part (and why you should not wait until the last minute). Read through your report one last time. AI is awesome, but it doesn't know your voice like you do. Make sure everything sounds like you wrote it, not like a robot. Add your own flair, your own jokes and your own personality.

As you're reading, think about the flow of your ideas. Does one paragraph lead smoothly into the next? If not, you might need to add some transition sentences. It is like building bridges between your ideas and hey, we know all about bridges now, right?

Why let it sit? This helps you clear your head. You can think about other things, get a good night's sleep, then revisit the report. If it still makes sense and you are proud to add your name to it, submit it. Show the world your awesomeness.

Forward Thinking

Still here? I am honored. I chose the Golden Gate Bridge as a well-known example/tourist attraction. The Golden Gate Bridge is #19 in *What Are the Top 10 Tourist Attractions in the USA.*[v]

Homework

Time for some fun **and** to put this knowledge to use. Like Mark says in his cybersecurity classes, "Life is not a spectator sport."

Brooklyn Bridge

Seeing that I've covered the Golden Gate Bridge extensively, may I suggest writing a report on #20, the Brooklyn Bridge? I have never seen it. Educate me. Here's your homework:

"Write 500-1,000 words on the Brooklyn Bridge. Tell me where it is, what it connects, what color it is, who built it, when it was built, how tall it is, how long it is, how many cars travel on it every day and how I would get to it from Pittsburgh, by car." Send your response to mark@cybersafetynet.net.

Pittsburgh's Bridges

I attended a conference in Pittsburgh, Pennsylvania a few years ago. I scheduled a free day to explore the city. Glad I did. I learned about Pittsburgh's three rivers and several bridges. May I suggest researching Pittsburgh's bridges? Here's your homework:

"Write 500-1,000 words on Pittsburgh's Bridges. Tell me how many are there. Tell me about the three major bridges, who built them, what rivers they cross, how many cars travel on them every day and how I would get to Pittsburgh from Chicago, by car." Send your response to mark@cybersafetynet.net.

Chicago's Bridge That Fell Up

I saved this one for last. I used to live in Chicago. One day, a bridge malfunctioned and fell up. I kid you not. Here's your homework:

"Write 500-1,000 words on Chicago's Bridges. Tell me how many are there. Tell me about the bridge that *fell up*, why it happened and how I would get to Chicago from Pittsburgh, by car." Send your response to mark@cybersafetynet.net.

Section Summary

Just remember, we're talking about AI as a research tool here. We knew nothing about the CIA when we entered that prompt. Only after we gained some information were we able to choose a) follow-up the initial answer, or b) go down a different path. I think you'll find yourself embracing the same strategy when researching the Brooklyn Bridge, Pittsburgh's bridges and Chicago's bridge that fell up. I am intentionally withholding the name here. I've found the more I research with AI, the better my initial prompts and hence, the stronger my follow-ups. Enjoy the journey.

Section Quiz

Life is not a spectator sport. Here's a quiz.

1. **What is the primary AI tool discussed in the text?**

a) ChatGPT
b) Perplexity
c) DALL-E
d) Google AI

2. **According to the text, what is the official term for asking questions to an AI?**

a) Inquiring
b) Querying
c) Prompting
d) Interrogating

3. **What feature of Perplexity allows users to explore different angles of a topic?**

a) Ask follow-up
b) Focus feature
c) Deep dive
d) Topic explorer

4. **What color is the Golden Gate Bridge?**
a) Red
b) International Orange
c) Golden
d) Silver

5. **In what order does the text suggest writing the sections of a report?**
a) Introduction, Body, Conclusion
b) Body, Introduction, Conclusion
c) Body, Conclusion, Introduction
d) Conclusion, Body, Introduction

6. **What does the text compare creating an outline to?**
a) Building a house
b) Cooking a meal
c) Drawing a map
d) Solving a puzzle

7. **What does the text suggest doing to check the flow of your writing?**
a) Ask a friend to read it
b) Use Perplexity to review it
c) Read it out loud
d) Wait a week before reviewing

8. **What is the purpose of citing sources in a report, according to the text?**
a) To make the report longer
b) To show you've done your homework
c) To confuse the reader
d) To impress the teacher

9. **What does the text recommend doing before final review of your report?**
a) Rewrite the entire report
b) Let it sit overnight
c) Have a friend proofread it

d) Submit it immediately

10. **What bridge does the text suggest as a homework topic?**

a) Golden Gate Bridge
b) Brooklyn Bridge
c) London Bridge
d) Sydney Harbour Bridge

Answers

1. b) Perplexity
2. c) Prompting
3. b) Focus feature
4. b) International Orange
5. c) Body, Conclusion, Introduction
6. c) Drawing a map
7. c) Read it out loud
8. b) To show you've done your homework
9. b) Let it sit overnight
10. b) Brooklyn Bridge

Mastering AI-Powered Music

Welcome to the AI Music Scene

How Suno Works Its Magic

Do you believe we're living in a time when computers create music from words? It's like something from a futuristic sci-fi movie, but it happens now. Let me introduce a new tool called Suno that's disrupting the lives of music makers everywhere.

Suno is like having a whole band inside your computer. With Suno, you don't need to know how to play an instrument or read sheet music. Even I create with Suno and the only musical talent I have is turning up the volume when a good song comes on the radio.

All you need is a great idea and a computer that plays sound. Make sure you're in a place where it is okay to make noise (I mean, music). I usually create my music when nobody else is home.

So, how does Suno turn your ideas into tunes? Suno uses "generative AI." This means it makes brand new things that have never existed before, like original songs. It's trained in tons of music. It knows what sounds decent and what doesn't.

The Power is in the Prompt

When you use Suno, you tell it what kind of song you want by typing in a description. Suno then uses what it knows about music to create a song that matches your description. Prompt for a loud classic rock song with two guitars? Suno gives you classic rock. Prompt for an instrumental piano ballad? Suno gives you a piano ballad. Prompt for acid rock so violent you can't understand the singer? Suno gives you acid rock.

Getting Started with Suno

Start by creating your account.

- Visit https://suno.ai
- Click "Sign Up"
- Choose from Discord, Google, or Microsoft login options.
- Complete the registration process
- Access your new Suno dashboard
- Remember your username and password

The Better the Prompt, the Better the Result

Whether simple or complex, it all starts with the prompt. These 10 elements are like the ingredients in your musical recipe. When you mix them just right, you get a tune that'll win everybody's respect

Genre

The genre is like the overall vibe of your song. It sets the stage for everything else. **An example is:** "Create a funky disco track with a modern twist." This tells the AI to blend the groovy beats of disco with some contemporary flavors. You might end up with something that sounds like Daft Punk meets the Bee Gees!

Mood

The mood is all about feelings. It's what your listeners will experience emotionally. **An example is:** "Compose a melancholic ballad that builds to a hopeful climax." You're asking for a song that starts off sad but ends on an uplifting note. Think of Adele's *Someone Like You* and Led Zeppelin's *Stairway to Heaven.* They start off blue but finish strong.

Instruments

Picking your instruments is like choosing the characters for your musical story. Each one brings its own flavor to the mix. **An example is:** "Use a combination of piano, acoustic guitar and subtle electronic beats." This prompt would create a song that blends organic and digital sounds. Think of Ed Sheeran's *Shape of You* or Air Supply's *All Out of Love.*

Tempo

Tempo is all about speed. It can make your listeners want to dance or relax. **An example is:** "Create a mid-tempo rock song at around 113 BPM (beats per minute)." This would give you a song with a steady, driving beat. Think of the Bee Gees' hit *Stayin' Alive*. Maybe you already knew this, maybe not, but the pulse of the average adult is 104 BPM. You may find yourself in a situation where you have to resuscitate somebody. Administer CPR, play *Stayin Alive* and measure the BPM. I am not making this up![vi]

Lyrics

Lyrics are the words that tell your story. They can be deep, fun, serious or silly, or anywhere in between. **An example is:** "Write lyrics about overcoming self-doubt and finding inner strength" This could result in an empowering anthem like Katy Perry's *Roar* or Lady Gaga's *Born This Way*.

Structure

Structure is like the blueprint of your song. It tells the AI how to arrange all the pieces. **An example is:** "Create a pop song with a verse-chorus-verse-chorus-bridge-chorus structure." This is a classic pop structure found in hits like Taylor Swift's *Shake It Off*.

Length

Don't have to be Albert Einstein for this one. How long do you want your tune to be? **An example is:** "Compose a 3-minute and 30-second song" This is a typical length for many pop songs, like Bruno Mars' *Just the Way You Are*. The Beatles' *I Want to Hold Your Hand*, *She Loves You* and *Yesterday* all run less than 2:30. Rush's *2112* (side one) completes in 20:33.

Inspiration

Inspiration is where you tell the AI to channel the vibes of other artists or songs. **An example is:** "Create a rock ballad inspired by Queen's *Bohemian Rhapsody* and Guns N' Roses' *November Rain*." This prompt would aim to capture the epic, theatrical nature of those classic rock ballads.

Era

The era helps set the time period vibe for your song. **An example is:** "Compose a song that sounds like it's from the 1980s." This could result in a track with big synths and drum machines. Think of Duran Duran and The Human League.

Special Effects

Special effects are the extra sprinkles on your musical sundae. They add that extra oomph! **An example is:** "Include a dramatic key change in the final chorus and end with a fade-out." This could create a powerful finale like what you hear in Whitney Houston's (Dolly Parton's) *I Will Always Love You*.

Remember, the more specific you are with these elements, the better the AI can understand your vision. It's like giving detailed instructions to your band members. The clearer you are, the closer the result will be to what you hear in your head. Onward and upward.

Assisted Prompts Energize Your Creativity

Jessie's Girl

Wait! Before you start writing prompts that end up as music, think twice about *who* will hear the music. You could end up in hot water.

If you create a song about wanting to steal your best friend's girlfriend, and your best friend finds out, you'll be in trouble. For further reference, see Rick Springfield's *Jessie's Girl* video.[vii] AI was not available at the time. Maybe Rick Springfield was 100% serious when he wrote and sang "You know I wish that I had Jessie's Girl."

Create Some Tunes

Ready to try your hand at creating some AI tunes? Yes, I can tell. Check out these sample prompts to get started:

- "Create an upbeat pop song with a catchy chorus about summer vacation. Use electric guitar, drums and synthesizers. Make it sound like something Taylor Swift would sing, with a length of 3 minutes."
- "Compose a mellow lo-fi hip hop track perfect for studying. Include soft piano, gentle beats and nature

sounds like rain in the background. The song should be 4 minutes long and have no lyrics."

- "Generate a high-energy EDM dance track with heavy bass drops and electronic synths. The lyrics should be about dancing all night. Make it 5 minutes long with a build-up in the middle leading to an epic drop."

Tunes I Created from Scratch

Can you believe a left brainer like me created music? My Suno screen name is *Zeus Lives in California*. Here are three of mine:

- With the prompt "A synthesizer-driven song to go with scuba diving video in the Caribbean. Slow, mysterious and refreshing. Mix in water sounds," I created two instrumentals, *Vacation (Time to Turn Off Your Phone)*[viii] and *Dive In (You Know You Want to be Here)*.[ix]
- With the prompt "A bright and slow techno pop song about viewing aurora borealis alone at midnight. Spatial, mysterious and galactic," I created two instrumentals, *Bright Midnight Lights*[x] and *Midnight Lights at My Piano*.[xi]
- With the prompt, "A loud, powerful rock n roll song about racing classic cars and one of the drivers crashes. Dual guitars. Strong male singer. Loud guitar solo," I created two rockers. The first one is *Thunder Road (Feel it in My Bones)*[xii] and *Twisted Metal Again on Thunder Road*.[xiii]

Tunes I Created from Suno's Ideas

I created the above prompts from scratch. If you have writer's block, consult Suno's ***Need ideas?*** Section. I see idea boxes for Person, Special Occasions, Moods, Activity, Greeting, Vacation, Scrapbook and Pet. You can create from one of these idea boxes. Here are some more songs I created:

From the Special Occasion Idea

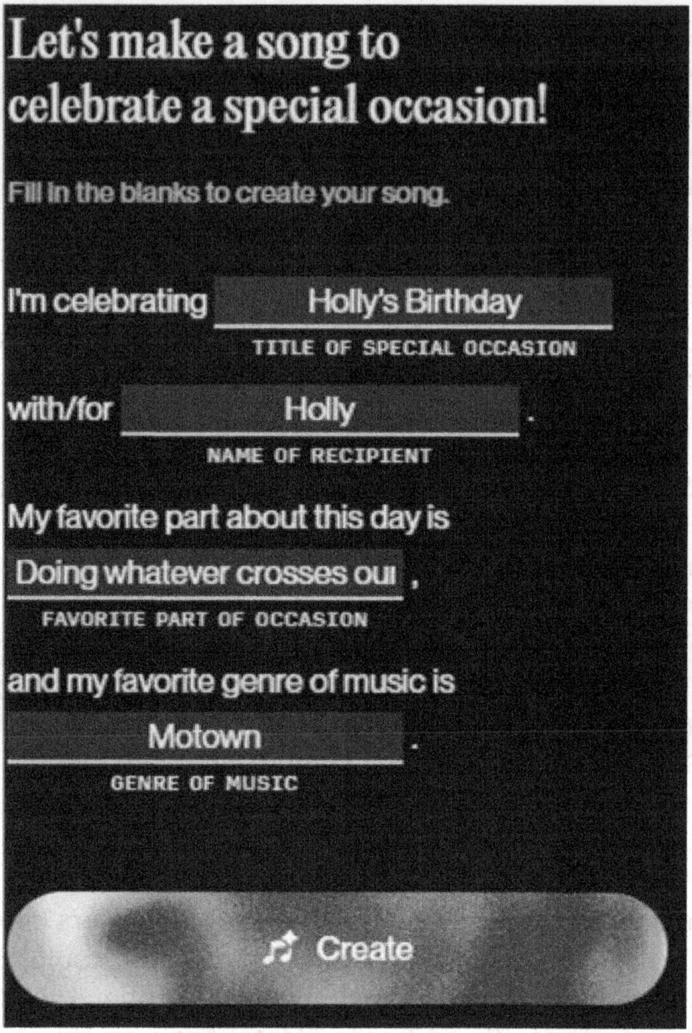

Original title: Holly's Birthday Bash
My version 1 title: *Whatever Joys*
My version 2 title: *Holly's Birthday Bash* (original)

From the Pet Idea

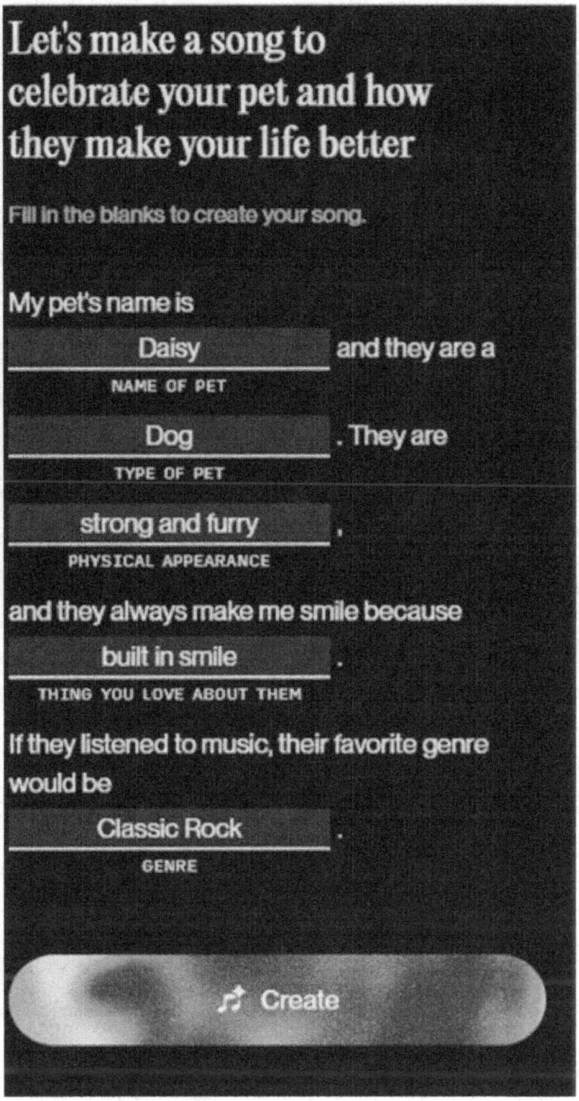

Original title: *Daisy's Anthem*
My version 1 title: *Wet Nose Nudging*
My version 2 title: *Daisy's Anthem* (original)
From the Moods Idea

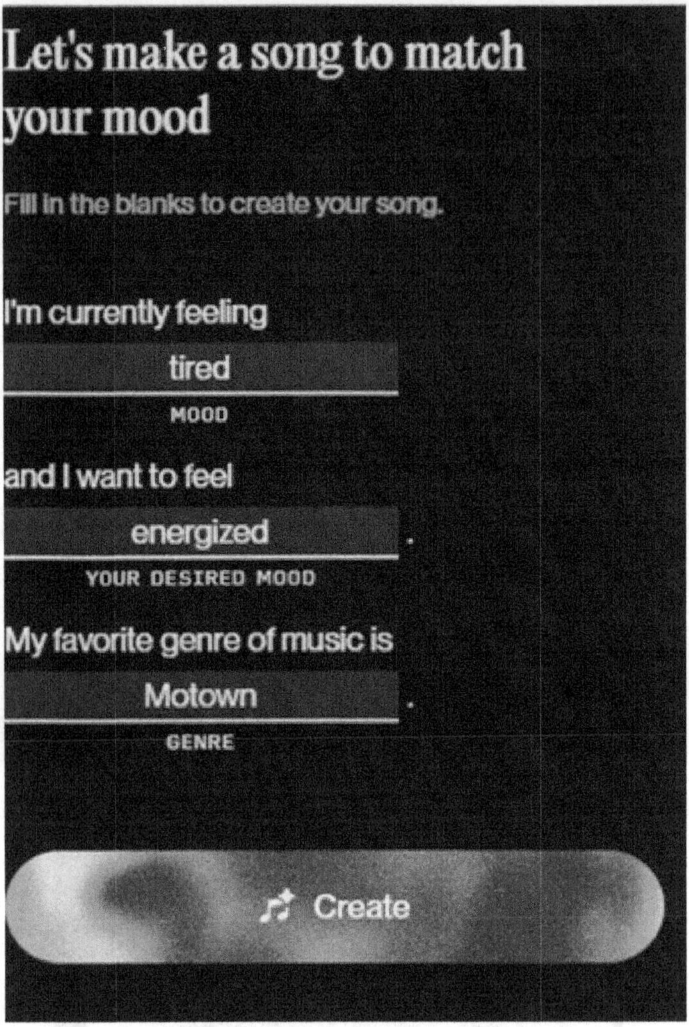

Original title: *Get Up and Dance*
My version 1 title: *Need Some Sunshine*
My version 2 title: *Get Up and Dance* (original)

Two at a Time?

Suno receives your prompt and generates two songs. Although you get 50 credits per day, each prompt creates two songs at 5 credits each. Let me clarify the math. Five prompts per day gets you 10 songs per day. Free account.

Playlists

I create new songs at least once a week. I am sharing them in a playlist titled *AI Ain't So tough: Wide variety of styles and themes portraying the tapestry of our lives. Accompanies the AI Ain't So Tough series at Amazon Kindle.*[xiv] Just bookmark that once and check back often. This playlist contains 257 songs as of the date of publication.

Is Any Music Original?

One of the world's most revered songs is Led Zeppelin's *Stairway to Heaven*. Jimmy Page and Robert Plant wrote *Stairway to Heaven*. Not everybody agrees with that statement.

The controversy over Led Zeppelin's *Stairway to Heaven* centers on claims that the band copied its famous opening guitar riff from *Taurus*, an instrumental track by the band Spirit. *Taurus* was written by Spirit's guitarist Randy Wolfe (known as Randy California) and released in 1968, three years before *Stairway to Heaven* debuted in 1971. Wolfe's estate filed a lawsuit in 2014, arguing that Led Zeppelin had access to *Taurus* because the two bands performed at the same venues during Zeppelin's early U.S. tours.

The lawsuit claimed Jimmy Page borrowed the descending chord progression from *Taurus*. Mark Andes, Spirit's bassist, testified that Robert Plant attended a Spirit concert, but Plant said he didn't remember due to a car accident. Page denied hearing *Taurus* before the lawsuit.

Music experts debated whether the songs were similar. The plaintiffs claimed they created the chord progression, while Zeppelin's defense said it was a common musical sequence used for centuries. The jury ruled in favor of Led Zeppelin, finding *Stairway to*

Heaven was not "intrinsically similar" to *Taurus*. Copyright law does not protect generic musical elements like chord progressions.

After appeals, the case ended in 2020 when the U.S. Supreme Court declined to hear it, upholding earlier rulings in Led Zeppelin's favor. The band avoided paying damages and retained full rights to their iconic song.

The Pros and Cons of AI-Generated Music

Like most things these days, nothing is 100% great while something else is 100% awful. There are pros and cons to everything.

Pros

- Anyone can create music, even if they can't play an instrument
- It's super-fast, therefore you can make a whole song in minutes
- You can experiment with different styles easily.
- It's a great tool for inspiration and coming up with new ideas

Cons

- Some people worry it might replace human musicians
- The music might not have the same emotional depth as human-created songs
- There are still questions about copyright and ownership
- It might make the music industry even more competitive

Ethical Considerations in AI Music

Let's take a moment to think about some of the big questions this technology raises:

- How can we make sure AI doesn't put human musicians out of work?
- Should AI-generated music be labeled as such when it's released?
- How can we ensure that AI music tools are accessible to everyone, not just big companies?

These are tough questions without easy answers. As you explore the world of AI music, keep thinking about these issues and how you can use this technology in a way that is fair and beneficial for everyone.

Maybe you knew this and maybe you didn't, but Pedro Sandoval made two albums entirely with AI.[xv]

AI's Impact on the Music Industry

AI isn't just changing how we create music. It is shaking up the whole music biz. Here's how:

- Music production is getting faster and cheaper
- Small artists can compete with big labels more easily
- New job opportunities are popping up for people who know how to use AI music tools
- The way we discover new music might change, with AI helping to match listeners with songs they'll like

Don't worry. Remember when I wrote about polishing and following-up prompts to get better results from Perplexity? You can polish AI-generated music to show human creativity and emotion in music. AI-generated music creates a starting point, but somebody has to polish it to make the music more appealing.

You cannot place blind faith in AI. Just because you ask Suno for a Motown song does not mean you get Motown. I have prompted Suno for Motown and female singers more than 12 times. Suno consistently fails this request. There are definite shortfalls in AI-created work. You **must** proofread, or in this case, listen, with a critical ear.

The Future of Music Creation

Alright music lovers (or by now, maybe I should say "music creators"), we are nearing the end of this section. As AI technology gets even smarter, we might see:

- AI that can create entire albums with connected themes
- Virtual AI bands that perform "live" concerts
- AI that can remix and mashup songs in totally new ways
- Tools that let you collaborate with AI to make music together

The most exciting possibilities might come from humans and AI working together to create awesome new sounds.

Homework

Time for some fun and to put this knowledge to use. Like Mark says in his cybersecurity classes, "Life is not a spectator sport."

Ready to create your own music? Yes, I am sure you are. Here are some tips to get you started:

- Experiment with different AI music tools. Suno is great, but there are others out there too.
- Practice writing good prompts. The more specific your prompts, the better results you'll get.
- Listen to lots of different kinds of music for inspiration.
- Don't be afraid to mix and match genres and styles.
- Share your creations with friends and get feedback.
- Remember that practice makes perfect. The more you use these tools, the better you'll get.

Section Summary

It is now your turn to rock the AI music scene. We've come to the end of our AI music journey. You've learned about Suno, how AI creates music, the power of prompts and the ambiguity of copyright and ethics.

Now it's your turn to get out there and start making some AI-assisted tunes! Remember, the most important thing is to have fun and let your creativity flow. Who knows? Maybe the next big hit will come from your collaboration with an AI!

Keep rocking, stay curious and never stop exploring the growing world of music.

Section Quiz

Life is not a spectator sport. Here's a quiz.

1. **What is Suno?**

a) A new type of musical instrument
b) An AI-powered music creator
c) A record label
d) A music streaming service

2. **What is a prompt in AI music creation?**

a) A type of musical note
b) A description of the song you want the AI to create
c) A music theory concept
d) A type of microphone

3. **Who, according to the U.S. Supreme Court, wrote *Stairway to Heaven*?**

a) Jimmy Page
b) Robert Plant
c) Randy California
d) Trick question, the U.S. Supreme Court refused to hear the case

4. **What is one advantage of AI-generated music?**
a) It always sounds better than human-made music
b) It's faster to create than traditional methods
c) It's more expensive
d) It can only create classical music

5. **What is one ethical concern about AI-generated music?**
a) It might replace human musicians
b) It can only create short songs
c) It requires expensive equipment
d) It can only be used by professional musicians

6. **What is generative AI?**
a) AI that can only analyze existing music
b) AI that can create new, original content
c) AI that can only play musical instruments
d) AI that can only write lyrics

7. **Which of these is NOT typically included in a music generation prompt?**
a) Genre
b) Mood
c) The creator's favorite color
d) Tempo

8. **How might AI impact the music industry?**
a) By making music production faster and cheaper
b) By eliminating all human musicians
c) By making all music sound exactly the same
d) By making music creation more difficult

9. **What is one way to improve your AI music creation skills?**
a) Never listen to human-made music
b) Only use one specific prompt
c) Practice writing different types of prompts
d) Avoid sharing your creations with others

10. **What is an important consideration when using AI-generated music?**
a) Ownership issues
b) The need to have perfect pitch
c) The requirement to own multiple instruments
d) The necessity of a music theory degree

Answers

1. b) An AI-powered music creator
2. b) A description of the song you want the AI to create
3. d) Trick question, the U.S. Supreme Court refused to hear the case
4. b) It's faster to create than traditional methods
5. a) It might replace human musicians
6. b) AI that can create new, original content
7. c) The creator's favorite color
8. a) By making music production faster and cheaper
9. c) Practice writing different types of prompts
10. a) Ownership issues

Rock on, music creators! How many did you get right? Remember, learning is a journey, not a destination. Keep practicing and exploring the emerging world of AI music creation!

Tilly Norwood

Say hello to Hollywood's newest sensation, Tilly Norwood. "Never heard of her," you may say. I understand.

Some love her and some hate her. She looks like she's in her mid to late 20s, but she is just 12 days old (as of this writing).

Tilly Norwood represents a significant development in the intersection between artificial intelligence and entertainment. Tilly is the first fully AI-generated actress to achieve international attention. She has become a subject of both professional admiration and controversy in the entertainment industry.

Popularity and Appeal

Tilly Norwood gained rapid popularity following her launch by Dutch producer Eline Van der Velden and the teams at Xicoia and Particle6. Her presence on social media platforms, including Instagram, quickly attracted attention due to her beautiful appearance and compelling digital performances. Tilly's appeal stems from the strategic combination of algorithmically perfected features and a relatable personality, modeled to evoke familiarity and charm. Industry observers often compare her presence and mannerisms to Scarlett Johansson and Natalie Portman. Her British accent and balanced public persona add a layer of authenticity that resonates with global audiences.

Creation and Concept

Tilly Norwood has no acting background or personal experience. She was created entirely through artificial intelligence, drawing upon advanced generative models for both visual and behavioral dimensions. Under Eline Van der Velden's leadership, Tilly was designed as a "work of art" that redefines the boundaries between technology and creative expression. The intent behind her development was to introduce new possibilities for storytelling, production efficiency, and digital artistry within the entertainment sector. According to her creator, Tilly is not intended to supplant human actors but to supplement and challenge traditional approaches to performance.

Support and Reception

The entertainment industry has noticed. Talent agencies and producers express interest in the efficiencies she offers. These include flexibility in casting, multilingual capabilities and cost-effectiveness. She can be in two places at once. She's not part of a union. She won't ruin a film's marketing by getting drunk and smashing a car into a tree. These traits make her a promising figure for international productions and experimental film projects. In addition, her popularity on social media demonstrates a growing audience receptive to digital performers.

Naysayers

However, not everybody loves her. The most vocal opposition comes from performers' unions such as SAG-AFTRA, prominent actors and industry stakeholders. Critics argue that the rise of AI actors risks undermining employment opportunities for human performers, especially those starting their careers. Concerns include issues of consent, intellectual property and the impact of AI creation on artistic integrity. Some audience members find Tilly's presence unsettling. They say she lacks genuine emotional expression. Give her a break. She's 12 days old!

Her Future is Bright

Tilly Norwood's popularity comes from being first. She is a test case for what AI can do in entertainment. Tilly quickly gathered over 40,000 Instagram fans, a fanbase that loves novelty, high-gloss visuals and the chance to be part of virtual history. Studios and talent scouts are excited in her economy: Tilly never asks for overtime, doesn't age and can be endlessly re-tooled for any language, style, or genre at a moment's notice. She is more versatile than any person.

Section Quiz

Tilly Norwood is new. Remember, she is 12 days old as of this writing. I cannot think of any quiz questions. Let's just go to the next section.

AI Disclosure

Ethical Considerations Expand Outside of Music

Remember the *Ethical Considerations in AI Music* section a few pages back? It is time to expand the ethical discussion outside of music and into AI-generated content in general.

Picture this: You're scrolling through your favorite social media app and see an image of a beautiful woman.

Your first thought is probably, "Whoa, that's awesome!" But then you might wonder, "Wait a sec... is she even real?"

This is where AI disclosure comes into play. It's here to keep reality from getting too artificial.

AI disclosure is like telling your reader, "Hey, I used some computer magic here!" It's all about being honest and letting people know when you used artificial intelligence to create or tweak something.

Deepfakes

Imagine you're watching a video of Barack Obama using profanity and saying some things that are out of character. You ask, "Did he really say that?" Someone posted a deepfake video,[xvi] of Brack Obama online. The language is harsh. The video is fake. This deepfake is one minute, 22 seconds long. You can find it on YouTube. Watch to the end to see what is *really* happening.

Thank you for watching that one to the end. The person publishing that video honestly showed what he was doing behind the scenes. He was masking his words in Barack Obama's likeness and voice. Bravo to the publisher for his disclosure.

Others are Not Honest

Fasten your seatbelt. Here are some more deepfakes. These, however, do not end with the creator showing how he did it. The intention is more fraudulent. Disclosure is not happening.

They don't have to be videos. Back during the election season of 2024, someone created a deepfake voice recording of President Joe Biden encouraging Democrats to **not** vote[xvii] in the New Hampshire primary.

They don't have to feature international celebrities. Someone created a deepfake audio recording of a Maryland school principal making derogatory, racist and antisemitic remarks. People believed it. The Baltimore Police Chief did not.[xviii] The suspect is the Athletic Director at the school where the principal was investigating him over an alleged $2,000 theft.

How Can You Tell It is Deepfake?

Yes, this is the $1,000,000 question. Let me share what my high school journalism teacher preached when evaluating a source.

- What is the source's agenda? After the source (allegedly) of the Maryland deepfake audio was discovered, police determined his motive.
- Do at least two other sources report the same news? Back when the Joe Biden deepfake happened, it was on the news as a story of deepfake scams, not a legitimate message from the President. News outlets were not fooled.
- Is it credible? Watch the lips. Does the video match the sound?

I am adding "Is this within character?" I have never seen the real Barack Obama swear at a camera. Keeping his cool is partly responsible for Barack Obama getting elected President twice.

Why Should You Care?

Now, you might be scratching your head and thinking, "But why should I care if AI was used?" Excellent question. People expect you to be honest. **Once you break their trust, you may never earn it back.**

To Believe or Not to Believe

When you know something was created or changed by AI, you can make better decisions about what to believe. It's like having x-ray vision that helps you spot the difference between what's real and what's just good computer wizardry. You could say it gives a clearer *pixelture* of the situation! Get it? Pixel... picture? Oh, come on, that was a good one! I don't crack jokes too often. Can I get some love here?

Why Bother?

- Being honest helps your reader trust you and respect your honesty
- You can appreciate the real skills of human artists and creators
- It makes readers think twice before sharing something that might not be true

The Detective Game

Alright, put on your detective hats because we're about to play a game I like to call *Spot the AI*. Think of it as a digital version of *Where's Waldo*? but instead of finding a guy in a striped shirt, we're looking for signs of artificial intelligence. It's time to put on your thinking caps and solve the mystery of AI honesty!

Here are some clues you might look for:

- Images that look too perfect or have weird details
- Videos where people's mouths don't quite match their words (this happens often, by the way)
- Text that sounds a bit robotic or does not quite make sense, like if I suddenly started talking about quantum physics in the middle of a sentence about ice cream

Start by searching for deepfake videos then move on to pictures, audio and text. When you find some unbelievable content, send me a quick email with the links.

Three-Legged Dancer

Let me start the fun with one that is so wrong it becomes amusing. Look for dancers with three legs.

By being on the lookout for these clues, you're not just protecting yourself. You're helping everyone stay informed. You cannot have blind faith in AI. You may end up with a picture or video so ridiculous that you end up embarrassing yourself when you share it.

Yes, this qualifies as an AI hallucination.

Fairness for Everyone

You know how back in school, everyone had to follow the same rules? Well, AI disclosure is kind of like that for the digital world. It makes sure everyone is playing fair and square, just like when you were picking teams for dodgeball. Don't want any cheaters in this tech tournament!

When people are honest about using AI, it means:

- Human artists don't have to compete with computers pretending to be humans
- Giving credit where it's due, to the AI creators and the human artists
- Everyone gets a fair shot at showing off their skills

The Art Gallery Adventure

Imagine you're visiting an art gallery. It has paintings made by humans and some made by AI. If they're all mixed together without labels, how would you know which is which? It might not seem fair to the human artists who spent hours perfecting their skills, right?

Now picture the same gallery, but this time, each piece has a little sign saying whether it was made by a human or AI. Suddenly, you can appreciate both types of art for what they are. You might think, "Wow, that AI can create great art" but also, "Woah, a human made that with their own hands? That's amazing!"

That's what AI disclosure does. It lets us appreciate both human creativity and technological innovation. It's like having the best of both worlds! You get to enjoy the Mona Lisa and the Robo-Lisa side by side. It's a real brush with greatness, whether it's made of bristles or pixels.

Understanding the Changing World

AI is changing the world faster than ever. By being open about when and how you use AI, we can keep up with these changes and understand them better. It's like we're all on this wild roller coaster of progress, and AI disclosure is our safety harness!

Knowledge is Power

Knowing about AI helps us:

- Figure out how it might affect our future jobs.
- Understand the disruptive new tech that's coming our way. The rate of change is astounding.
- Make smart choices about using AI in our own lives. This includes knowing when to use spellcheck and when to trust your head. Remember, there **is** a difference between "there," "their" and "they're."

Think of AI as your personal time machine. It gives you a sneak peek into the future! By understanding how AI is being used today, you can get a pretty good idea of what tomorrow might look like.

Do not surrender to any technology. You could end up with a three-legged dance partner!

The Good, The Bad and The AI Wild West Show

Now, let's talk about some of the awesome things that can happen when we're honest about using AI, and some of the not-so-cool stuff that can go down if we're not. It's like the Wild West of technology out there! So, saddle up, partners, and let's explore this digital frontier!

The Upside

When you're upfront about AI:

- People trust technology more because they know what's going on and impacting them.
- You can have interesting conversations about how AI is changing the world. This is more fun than talking about the weather, that's for sure.

- It's easier to spot and fix any problems with AI before they get out of hand. Let's correct the three-legged dancers out there.

The Downside

When we don't disclose we're using AI:

- Some folks might get tricked by fake news and make bad decisions, like believing that the Earth is flat.
- Real human artists might lose out to AI that's pretending to be human.
- We might miss out on understanding how AI is changing things around us, and eventually deciding humans are no longer necessary.

Writers and Artists are Nervous

Will AI take jobs? The Writers Guild of America (WGA) feels that can happen. They went on strike from May to September 2023. One of their three primary concerns was the role of AI in the entertainment industry. Folks who write movies and TV shows wanted artificial intelligence used only as a tool that can help research or facilitate script ideas. Studios felt AI could write scripts and then human proofreaders could polish them.[xix] Maybe the writers were overreacting. Maybe their views are valid. The strike has been settled, but AI's growing impact on the job market is a developing story.

Trusting Technology

To Trust or Not to Trust

Think about your best friend. You trust them because they're always honest with you, right? Well, AI disclosure is like that, but for technology. When companies and creators are honest about using AI, we start to trust them more.

It's like your friend told you they used a calculator to solve a math problem. You'd still think they're smart for knowing how to use the calculator, right? Same thing with AI. You can appreciate the skills of people who know how to use AI tools effectively.

Honesty is always the best policy, whether you're dealing with friends, robots, or anything in between.

Put AI Disclosure Advice to Good Use

Time to stop spectating and start creating. Here's your mission, should you choose to accept it:

- Always disclose when you've used AI to create or change something
- Ask questions when you're not sure if something was made by AI
- Share what you learn about AI with your friends and family

Your Toolkit

Here are some smart ways to disclose AI usage. You can a little note to your social media posts that says, "Created with AI" or "AI-enhanced." If you're making videos, mention in the description that you used AI tools. On your website, add a note in the footer. For example, www.cybersafetynet.net has this in the footer, "Full Disclosure: Some content here is generated by AI. The views expressed are the author's opinion and not legal advice. The author is not a lawyer. You are an adult and responsible for anything you do."

Practice What I Preach

Yes. I have included AI content in every book I've published since 2022. That includes this one. Let me disclose, "Some content here is generated by AI. The views expressed are the author's opinion and not legal advice. The author is not a lawyer. You are an adult and responsible for anything you do."

The Ripple Effect

When you're honest about using AI, you're starting a ripple effect of honesty. It's like throwing a pebble into a pond. Your small action can create waves that spread far and wide.

Here's how it works:

- You disclose your AI use (splash!)
- Your friends see it and think, "Hey, that's pretty honest and cool" (ripple...)
- They start doing it too (bigger ripple!)
- More and more people join in (wave alert!)
- Before you know it, everyone's upfront about AI (tsunami of honesty!)

You're not just following the rules... you're setting the trend. You're the new rock star.

Section Summary

As we wrap up our journey through the land of AI disclosure, remember that being honest about AI isn't just the right thing to do. It also shows that you are smart, honest and ready for the amazing future that's coming our way. You're not just keeping up with the Joneses, you're keeping up with the Jetsons! Remember:

- Be curious about AI. Ask questions like they're going out of style.
- Be honest when you use it. Honesty is the best policy.
- Be excited about the future!

Mark has a brain teaser. Three people are standing at a train station. The first person says, "The train is coming." The second person says, "The train is here." The third person says, "What train?" Which person do you want to be?

Section Quiz

Life is not a spectator sport. Here's a quiz.

1. **What is AI disclosure primarily about?**
a) Hiding the use of AI
b) Promoting AI technology
c) Being honest about using AI in content creation
d) Criticizing AI-generated content

2. **Which of the following is NOT mentioned as a way to identify a deepfake?**

a) Checking the source's agenda
b) Verifying if multiple sources report the same news
c) Assessing if the content is within character for the person depicted
d) Examining the file metadata

3. **Why is AI disclosure important for human artists?**
a) It helps them learn AI techniques
b) It prevents unfair competition with AI-generated art
c) It increases the value of AI-generated art
d) It makes human art obsolete

4. **What analogy does the text use to explain the concept** of AI disclosure in an art gallery?
a) A time machine
b) A roller coaster
c) Labels on artwork indicating human or AI creation
d) A game of dodgeball

5. **According to the text, how can AI disclosure help us understand the changing world?**
a) By predicting exact future events
b) By providing insights into potential future jobs and technology
c) By replacing human decision-making
d) By slowing down technological progress

6. **What is the "ripple effect" of honesty mentioned in the text?**
a) It causes water pollution
b) It leads to widespread dishonesty
c) It encourages others to also be honest about AI use
d) It creates waves in the ocean

7. **Which of these is NOT suggested as a way to disclose AI use?**
a) Adding a note to social media posts
b) Mentioning AI use in video descriptions

c) Including a disclosure in a website footer
d) Tattooing "AI user" on your forehead

8. **What potential negative consequence of not disclosing AI use is mentioned in the text?**
a) Artists going on strike
b) People might assume fake news is real
c) Computers taking over the world
d) The end of human creativity

9. **How does the text suggest AI disclosure can impact trust in technology?**
a) It decreases trust in all technology
b) It has no impact on trust
c) It increases trust by promoting transparency
d) It makes people overly reliant on AI

10. **What metaphor does the text use to describe the current state of AI and disclosure?**
a) The Wild West
b) A school playground
c) A scientific laboratory
d) An ancient civilization

Answers

1. c) Being honest about using AI in content creation
2. d) Examining the file metadata
3. b) It prevents unfair competition with AI-generated art
4. c) Labels on artwork indicating human or AI creation
5. b) By providing insights into potential future jobs and technology
6. c) It encourages others to also be honest about AI use
7. d) Tattooing "AI user" on your forehead
8. b) People might assume fake news is real
9. c) It increases trust by promoting transparency
10. a) The Wild West

What the Lawyers and Courts Say

Introduction to AI and Copyright

Maybe you have an opinion on owning or not owning the content you generate. Lawyers and courts have opinions too. Their opinions speak louder than yours and mine. While the Supreme Court hasn't made any major decisions about AI and copyright yet, lower courts have.

What is Copyright?

Copyright is a law that protects original works like books, movies, songs, paintings and software. It gives the creator of a work the right to decide how it's used. For example:

- If you write a book, no one else can copy or sell it without your permission.
- If you paint a picture, you can decide who gets to print or display it.
- If you record a song you wrote, no one else can play it at a public event without your permission. This is why some artists complain when political campaigns play their songs at political rallies.

For something to be copyrighted in the United States, it usually must meet two big rules:

- It must be original: This means it must be new (and not a copy).
- It must be made by a human: This is a grey area. Shouldn't be, but it is.

Even though the US Supreme Court has not ruled on AI yet, lower courts have. Here are three examples from Perplexity:

Case 1: Art Created by AI Cannot Be Copyrighted

Background

One of the most important rulings happened in 2023 when a federal court decided that art created *entirely* by AI cannot be copyrighted. This case involved Stephen Thaler, who used an AI program called "Creativity Machine" to create a piece of art called *A Recent Entrance to Paradise*. Thaler tried to register the artwork for copyright protection and was denied.

What the court said

In *Thaler v. Perlmutter*, the court said that U.S. copyright law requires human authorship. In other words, only humans can own copyrights. The judge explained that while AI can assist in creating art, it cannot replace the human element needed for copyright protection.

This decision was a big deal because it set a clear rule. If an artwork is made entirely by AI with no human input, it cannot earn copyright.

Mr. Thaler was not amused. He appealed to the U.S. Court of Appeals for the District of Columbia Circuit (Washington, D.C.). The appeals court upheld the lower court's verdict on Tuesday, March 18, 2025 (yesterday. I kid you not. I am writing this section on the 19th. Perfect timing, eh?).

What we learned

Only humans can get copyrights in the U.S. This means if an artist uses an AI tool like DALL-E or MidJourney to help create a painting or digital artwork, they can get copyright protection *only* if they play an active role in making creative decisions.

For example:

- If you use an AI program to generate random images but don't do anything else, those images probably won't be protected by copyright.
- But if you use an AI tool as part of your creative process, like editing or combining images, the final work might qualify for copyright because you added your own creativity.
- Music is art too. In the *Mastering AI Powered Music* section, I show how to generate songs with Suno. I cannot claim copyright on any of my songs. I just typed a prompt, pressed (Enter) and published what Suno generated.

You can find this art online.[xx]

Case 2: Using Copyrighted Material to Train AI

Background

Another important case involved ROSS Intelligence, a company that used AI to build a legal research tool. ROSS trained its AI system with data from Westlaw, which is a legal database that Thomson Reuters owns. Thomson Reuters sued ROSS for copyright infringement, arguing that ROSS had illegally used its copyrighted material.

What the court said

In *Thomson Reuters v. ROSS*, the court ruled against ROSS Intelligence. It said that using copyrighted material to train an AI system without permission could be considered copyright infringement. ROSS argued that its use of Westlaw's data was "fair use," which is a legal defense that allows limited use of copyrighted material in certain situations (like education or research) if they transform the original work into something new and don't harm the original creator's business. However, the court rejected this argument because ROSS was using the data to create a competing product.

How recent is the Ross Intelligence case? It was heard in the U.S. District Court for the District of Delaware. The initial lawsuit was filed in May 2020. Significant rulings occurred on September 25, 2023, and again on February 11, 2025 (not as recent as yesterday, but still recent).

This case showed that companies need to be very careful about how they train their AI systems. Using copyrighted material without permission could lead to lawsuits. You make enemies if you harvest a company's data and use it to make a competing product. Somebody at ROSS should have thought twice and yelled "BAD IDEA! BAD IDEA!" Nuf said.

What we learned

The fair use argument might not work. Using someone else's copyrighted material without permission is risky, especially if you're building a competing product. Courts are becoming stricter about what counts as "fair use," especially when it comes to creating products that compete with existing ones.

This ruling could make it harder for companies and researchers to train their AI systems unless they have permission from copyright owners. It also raises questions about whether we need new laws, now that AI is so prominent.

Case 3: Harvesting Free Data

Background

The New York Times and OpenAI are in a heated legal battle over how OpenAI uses the newspaper's content to train its AI models. This includes ChatGPT. The disagreement revolves around whether OpenAI violated copyright laws by harvesting (copying) articles from the Times without permission or payment. This case is important because it could change how AI companies interact with copyrighted materials and how media organizations protect their content.

Problem? What Problem?

The New York Times claims that OpenAI harvested millions of its articles to train ChatGPT without asking for permission or paying licensing fees. When ChatGPT answers questions, it sometimes uses information from the Times' articles, and in some cases, it repeats parts of those articles word-for-word. The New York Times argues this is unfair and illegal.

OpenAI argues that its use of the articles is protected by "fair use," which I already covered. OpenAI says its AI transforms the articles into new outputs and doesn't directly compete with the Times.

The New York Times' Arguments

The New York Times believes OpenAI's actions are harmful for several reasons:

- **Copyright Infringement**. The Times says OpenAI copied its articles without permission, which breaks copyright laws. It also claims that ChatGPT sometimes repeats parts of its articles exactly as they were written, which is called "regurgitation." This means OpenAI isn't learning from the content. OpenAI is copying it.
- **Hurting Business**. The Times argues that ChatGPT competes with its business by giving people answers based on its reporting without sending them to its website. Normally, when people read an article online, they might see ads or pay for a subscription. But if ChatGPT gives them all the information for free, fewer people visit the Times' website. They lose revenue.
- **Evading Paywalls**. Some users have found ways to use AI tools like ChatGPT to bypass paywalls on websites like the Times'. This means people can access paid content for free using AI products, which further damages the newspaper's ability to make money.
- **Time-Sensitive Content**. The Times says OpenAI uses content from Wirecutter (its product review site) without proper credit. For example, during Black Friday sales,

Wirecutter publishes recommendations for deals on products. If ChatGPT uses this information without linking back to Wirecutter, it takes away potential earnings from affiliate links and damages Wirecutter's reputation.

- **Billions in Damages**. The Times claims that OpenAI's actions could cost it billions of dollars in damages because so many articles were used without permission.

OpenAI's Arguments

OpenAI disagrees with the accusations and argues that what it does is legal:

- **Fair Use**. OpenAI says its use of the Times' articles follows fair use rules because it transforms the original content into something new when training its AI models. For example, instead of copying an article word-for-word, ChatGPT learns patterns and uses them to create unique answers.
- **Rare Bugs**. OpenAI admits that ChatGPT sometimes repeats text exactly as it appears in an article but calls this a "rare bug" that they are working to fix. They also say some examples provided by the Times were intentionally manipulated to make ChatGPT repeat content.
- **No Proof of Harm**. OpenAI argues that the Times hasn't proven how it was harmed by these actions. For example, even if ChatGPT removes identifying information like author names or mastheads (copyright management information), OpenAI says this doesn't necessarily hurt the newspaper's business.
- **Negotiation Efforts**. Before the lawsuit started, OpenAI says it tried to negotiate a licensing deal with the Times but couldn't reach an agreement. OpenAI has already made deals with other media companies like the

Associated Press and Axel Springer to pay for their content.

Why Does This Matter?

This case is important because it could change how AI companies train their models and how media organizations protect their content:

- **Impact on AI Companies**. If OpenAI loses this case, it might have to pay billions of dollars in damages and stop harvesting copyrighted content without permission. This could make it harder for AI companies to train their models using existing data and might lead to stricter rules about what data it can use.
- **Impact on Media Organizations**. If the New York Times wins, it could set a precedent that forces AI companies to pay licensing fees whenever they use copyrighted material for training purposes. This would give media organizations more control over their content and help them earn money from AI companies.
- **Balancing Innovation and Copyright**. The case raises big questions about how we balance innovation with protecting intellectual property rights. Should AI companies be allowed to use copyrighted material freely if it leads to new technology? Or should creators be paid for their work no matter what?

What Could Happen Next?

The court hasn't made a final decision yet, but here are some possible outcomes:

- **OpenAI Wins**. If OpenAI wins, it might set a precedent that allows AI companies to use copyrighted material under fair use rules without paying licensing fees.
- **The New York Times Wins**. If the Times wins, AI companies might have to pay large amounts of money

for using copyrighted material in their training datasets or stop using such material altogether.

- **Settling Out of Court**. Both sides might decide to settle out of court by agreeing on a payment amount or licensing deal before the court makes a final ruling.

Stay tuned!

The OpenAI vs. New York Times case has the official name *The New York Times Company v. Microsoft Corporation et al* and is being heard in the U.S. District Court for the Southern District of New York. The lawsuit was filed on December 27, 2023

Will the Supreme Court Get Involved?

That is a safe bet. So far, most of these cases have been decided by lower courts or appeals courts. As more cases involving AI come up, the Supreme Court will eventually weigh in. If it does, its rulings could set nationwide laws about how copyright applies to AI-generated works.

Section Quiz

Life is not a spectator sport. Here's a quiz.

1. **What is required for a work to be copyrighted in the United States?**
   a) It must be created using AI tools.
   b) It must be original and made by a human.
   c) It must be publicly displayed.
   d) It must be registered with the Supreme Court.

2. **What was the key ruling in the case of Thaler v. Perlmutter regarding AI-generated art?**
   a) AI-generated art can be copyrighted if it is original.
   b) AI-generated art cannot be copyrighted unless humans are actively involved in its creation.
   c) AI-generated art automatically qualifies for copyright protection.
   d) AI-generated art can only be copyrighted if it competes with human-created works.

3. **Why did Thomson Reuters sue ROSS Intelligence?**
   a) ROSS Intelligence used Westlaw's data without permission to train its AI system.
   b) ROSS Intelligence copied articles from the New York Times for training purposes.
   c) ROSS Intelligence created songs using AI without proper licensing.
   d) ROSS Intelligence used its own proprietary data to train its AI system.

4. **What was the court's decision in Thomson Reuters v. ROSS regarding "fair use"?**
   a) The court accepted the fair use argument because ROSS transformed the data significantly.
   b) The court rejected the fair use argument because ROSS created a competing product using Westlaw's data.
   c) The court ruled that fair use applies to all AI training datasets without restrictions.
   d) The court did not address fair use in this case.

5. **What is one of the New York Times' main arguments against OpenAI in their legal battle?**
a) OpenAI transformed articles into new outputs, which is harmful to innovation.
b) OpenAI copied articles without permission, violating copyright laws and causing financial harm.
c) OpenAI negotiated licensing agreements with other media companies but excluded the Times intentionally.
d) OpenAI used outdated articles that were no longer relevant to current events.

6. **Why does OpenAI argue that its use of New York Times articles is protected under "fair use"?**
a) OpenAI claims it transforms original content into new outputs during training, rather than copying it directly.
b) OpenAI states that it only uses articles for educational purposes, not commercial gain.
c) OpenAI argues that it pays licensing fees for all articles used in training datasets.
d) OpenAI believes that copyright laws do not apply to AI models trained on public data.

7. **What potential impact could the New York Times v. OpenAI case have on media organizations?**
a) Media organizations may lose control over their content entirely if OpenAI wins the case.
b) Media organizations could gain more control over their content and earn licensing fees
c) Media organizations might have to pay damages to AI companies for restricting access to content used for training purposes.
d) Media organizations will likely stop publishing content online altogether due to copyright concerns.

8. **Which court upheld the decision in Thaler v. Perlmutter regarding human authorship?**
a) U.S. Supreme Court
b) U.S. District Court for Delaware
c) U.S Court of Appeals for the District of Columbia
d) Southern District Court of New York

9. **What does "regurgitation" refer to in the New York Times' argument against OpenAI?**
a) ChatGPT transforming articles into summaries without permission from authors.
b) ChatGPT repeating parts of articles exactly as they were written, rather than learning from them creatively.
c) ChatGPT bypassing paywalls and accessing restricted content illegally.
d) ChatGPT creating competing products using proprietary data from other companies like Westlaw or Wirecutter.

10. **Mark can copyright the music he created at Suno**
a) Yes, if he does not change the title
b) Yes, if he does not change the music
c) Yes, if he does not change the meta data
d) No

Answers

1. b) It must be original and made by a human
2. b) AI-generated art cannot be copyrighted unless humans are actively involved in its creation
3. a) ROSS Intelligence used Westlaw's data without permission to train its AI system
4. b) The court rejected the fair use argument because ROSS created a competing product using Westlaw's data
5. b) OpenAI copied articles without permission, violating copyright laws and causing financial harm

6. a) OpenAI claims it transforms original content into new outputs during training, rather than copying it directly
7. b) Media organizations could gain more control over their content and earn licensing fees
8. c) U.S Court of Appeals for the District of Columbia
9. b) ChatGPT repeating parts of articles exactly as they were written, rather than learning from them creatively
10. d) No

Case Study: 4-Day Work Week

Background

I've given you a lot of homework. How about I show you how prompting works for a highly controversial item these days, the 4-day workweek.

I am copying the thread from Perplexity with no content editing.[xxi] Sometimes the formatting is off. I am still keeping it as-is. Watch the follow-ups. My prompts are in quotes.

There will be 10 review questions at the end. Enjoy.

Why Adopt a 4-day Workweek?

"Tell me why my company should adopt a 4-day workweek."

Adopting a 4-day workweek can bring numerous benefits to your company, making it a compelling option for modern businesses. Here are several reasons why your company should consider implementing this innovative work schedule:

Increased Productivity

One of the most significant advantages of a 4-day workweek is the potential for increased productivity. Contrary to what you might expect, reducing work hours doesn't necessarily lead to less output. In fact:

- Microsoft Japan reported a remarkable 40% increase in productivity after implementing a 4-day workweek.
- Perpetual Guardian, a New Zealand company, saw a 20% spike in employee productivity.
- Buffer, a social media management company, found that 91% of their employees reported increased happiness and productivity

https://www.peoplebox.ai/blog/implementing-4-day-work-week-strategies/.

The increased productivity stems from employees being more focused and efficient during their work hours, knowing they have an extra day off to look forward to.

Improved Employee Well-being

A 4-day workweek can significantly enhance employee well-being:

- It promotes a better work-life balance, giving employees more time for personal pursuits and family.
- A study by the University of Cambridge found that a 4-day workweek led to a 71% reduction in burnout, a 39% decrease in stress, and a 65% reduction in sick days.
- 82% of companies reported positive impacts on staff well-being after implementing a 4-day workweek https://abcnews.go.com/Business/4-day-workweek-work-companies-share-results-after/story?id=107635577.

Happier, healthier employees are more likely to be engaged and committed to their work, which can lead to better overall performance.

Enhanced Recruitment and Retention

In today's competitive job market, offering a 4-day workweek can be a powerful tool for attracting and retaining top talent:

- 50% of companies reported reduced staff turnover after implementing a 4-day workweek https://abcnews.go.com/Business/4-day-workweek-work-companies-share-results-after/story?id=107635577.
- 32% of companies saw improvements in recruitment.
- 79% of employees consider flexible work schedules an extremely important benefit.

By offering this attractive perk, your company can stand out from competitors and become an employer of choice.

Cost Savings

A 4-day workweek can lead to significant cost savings for both the company and employees:

- Reduced office operating costs, including utilities and daily workplace perks.
- Lower commuting expenses for employees.
- Potential for up to 20% savings on variable overhead expenses.

These savings can be reinvested in the business or used to improve employee benefits.

Environmental Benefits

Adopting a 4-day workweek can also contribute to your company's sustainability efforts:

- Reduced commuting leads to lower carbon emissions.
- A study in the UK suggested that a 4-day workweek could shrink the country's carbon footprint by 21.3%

https://www.getclockwise.com/blog/four-day-work-week-benefits.

This aligns with growing consumer and employee expectations for environmentally responsible business practices.

Improved Work Quality

With more time to rest and recharge, employees often return to work more focused and creative:

- Companies like Buffer and Nectafy reported higher quality work and increased team trust after implementing a 4-day workweek.
- The extra day off allows employees to handle personal matters, reducing distractions during work hours.

In conclusion, adopting a 4-day workweek can lead to increased productivity, improved employee well-being, better talent attraction and retention, cost savings, environmental benefits, and higher quality work. While it may require some adjustment, the potential benefits make it a strategy worth considering for forward-thinking companies aiming to thrive in the modern business landscape.

--

"The reader is the CEO of a company that makes $100,000,000 a year and barely makes a profit. The firm has 10,000 full-time employees in the US. Write this response to present the concept, present three reasons that justify this policy change, present the opposing argument, then show why the argument in favor is stronger."

As the CEO of a large company with 10,000 full-time employees and $100 million in annual revenue, adopting a 4-day workweek could be a transformative strategy to boost productivity, improve employee satisfaction, and potentially increase profitability. While this change may seem radical, evidence suggests it could yield significant benefits for your organization.

Three Key Reasons to Adopt a 4-Day Workweek

1. Increased Productivity and Efficiency. Contrary to what you might expect, reducing work hours doesn't necessarily lead to less output. In fact, it often results in increased productivity:

- Microsoft Japan reported a remarkable 40% increase in productivity after implementing a 4-day workweek.
- Perpetual Guardian, a New Zealand company, saw a 20% spike in employee productivity.
- With fewer distractions and more time for deep work, employees often produce higher quality work in less time.

By focusing work hours and eliminating unnecessary meetings and tasks, your company could potentially see significant gains in productivity and efficiency.

2. Improved Employee Well-being and Retention

A 4-day workweek can significantly enhance employee well-being, leading to better retention rates:

- A study by the University of Cambridge found that a 4-day workweek led to a 71% reduction in burnout and a 39% decrease in stress.
- 50% of companies reported reduced staff turnover after implementing a 4-day workweek.

- Employees with better work-life balance are more likely to be engaged and committed to their work.

Given that turnover can cost a company up to two times an employee's salary, improved retention could lead to substantial cost savings.

3. Cost Savings and Operational Efficiencies

Implementing a 4-day workweek can lead to significant cost savings:

- Reduced office operating costs, including utilities and daily workplace perks.
- Potential for up to 20% savings on variable overhead expenses.
- Lower absenteeism rates, which can reduce associated costs.

These savings could help improve your company's profit margins without sacrificing productivity or employee satisfaction.

The Opposing Argument

Critics of the 4-day workweek argue that it may lead to increased costs and operational challenges:

- Some industries require 24/7 presence, making a 4-day workweek difficult to implement.
- There might be initial costs associated with restructuring work processes and potentially hiring additional staff to cover all necessary hours.
- Some employees may end up working the same hours anyway, potentially leading to overtime costs.

Why the Argument in Favor is Stronger

Despite these concerns, the potential benefits of a 4-day workweek outweigh the drawbacks:

1. The productivity gains and improved work quality could more than offset any potential increase in costs.
2. The reduction in turnover and associated costs could lead to significant long-term savings.
3. Improved employee well-being and satisfaction could result in a more engaged and loyal workforce, potentially leading to better customer service and innovation.

Furthermore, as a large company, you have the resources to carefully plan and implement this change, addressing potential challenges proactively. By offering a 4-day workweek, you could position your company as an employer of choice, attracting top talent and staying ahead of workforce trends.

In conclusion, while implementing a 4-day workweek requires careful consideration and planning, the potential benefits in terms of productivity, employee satisfaction, and cost savings make it a compelling option for your company. As CEO, leading this change could set your organization apart in a competitive business landscape and potentially drive significant improvements in both employee well-being and company performance.

--

"Write this as a speech the CEO would present to the Board of Directors."

Ladies and gentlemen of the Board, thank you for your time today. I'd like to present a bold initiative that I believe could transform our company's performance and position us as a leader in workplace innovation.

I'm proposing that we adopt a 4-day workweek for our 10,000 full-time employees across the United States.

Now, I know what you're thinking. With our annual revenue of $100 million and razor-thin profit margins, how can we afford to reduce our work hours? But I assure you, the evidence suggests this move could be the key to unlocking greater productivity, efficiency, and ultimately, profitability.

Let me outline three compelling reasons why this change makes sense for our company:

First, increased productivity and efficiency. You might expect productivity to drop with fewer work hours, but the opposite often occurs. Microsoft Japan saw a staggering 40% increase in productivity after implementing a 4-day workweek. Perpetual Guardian in New Zealand reported a 20% boost. By focusing our work hours and eliminating unnecessary tasks, we could see significant gains in output and efficiency.

Second, improved employee well-being and retention. A study by the University of Cambridge found that a 4-day workweek led to a 71% reduction in burnout and a 39% decrease in stress. Moreover, 50% of companies reported reduced staff turnover after this change. Given that turnover can cost up to two times an employee's salary, improved retention could lead to substantial savings.

Third, cost savings and operational efficiencies. We could see reduced office operating costs, including utilities and daily workplace perks. Some companies have reported up to 20% savings on variable overhead expenses. Lower absenteeism rates could further reduce costs. These savings could significantly improve our profit margins without sacrificing productivity or employee satisfaction.

Now, I acknowledge there are potential challenges. Critics argue that some industries require 24/7 presence, making a 4-day workweek difficult. There might be initial costs in restructuring work processes. Some employees might end up working the same hours, potentially leading to overtime costs.

However, I firmly believe the benefits outweigh these concerns. The productivity gains and improved work quality could more than offset any potential increase in costs. The reduction in turnover and associated expenses could lead to significant long-term savings. And improved employee well-being could result in a more engaged workforce, potentially driving better customer service and innovation.

As a large company, we have the resources to carefully plan and implement this change, addressing challenges proactively. By offering a 4-day workweek, we could position ourselves as an employer of choice, attracting top talent and staying ahead of workforce trends.

In conclusion, while implementing a 4-day workweek requires careful consideration, the potential benefits in terms of productivity, employee satisfaction, and cost savings make it a compelling option for our company. This change could set us apart in a competitive business landscape and potentially drive significant improvements in both employee well-being and company performance.

I'm open to your thoughts and questions as we consider this transformative initiative. Thank you.

--

"The Board of Directors approves the proposal. Write a press release announcing the new policy beginning at the start of the next fiscal year and add verbiage to satisfy the firm's doubters and disgruntled shareholders and convince them that this will benefit the firm's profits."

FOR IMMEDIATE RELEASE

[Company Name] Announces Transition to 4-Day Workweek, Aiming to Boost Productivity and Profitability

[City, State] - [Date]

[Company Name], a leading [industry] company with 10,000 employees across the United States, today announced its plans to implement a 4-day workweek for all full-time employees, effective from the start of the next fiscal year on [Date]. This bold move is designed to enhance productivity, improve employee well-being, and ultimately drive increased profitability.

"After careful consideration and extensive research, we are confident that this innovative approach to work will not only benefit our employees but also significantly contribute to our bottom line," said [CEO Name], CEO of [Company Name]. "We believe this change will position us as an industry leader in workplace innovation and help us attract and retain top talent in an increasingly competitive market."

Key points of the new policy include:

- Employees will work four 8-hour days per week, with no reduction in pay
- The company expects to see increased productivity and efficiency, based on successful implementations by other major corporations
- Projected cost savings from reduced operational expenses and lower turnover rates
- A phased implementation approach to ensure smooth transition and minimal disruption to operations

[Company Name] acknowledges that this decision may raise questions among shareholders and industry observers. However, the company is confident that this move will yield positive results:

1. Increased Productivity: Studies have shown that companies implementing a 4-day workweek have experienced productivity gains of up to 40%. [Company Name] expects similar improvements, which will directly impact our output and revenue.
2. Cost Savings: Reduced operational costs, including utilities and overhead expenses, are projected to result in significant savings. These savings will be reinvested in the company to drive growth and innovation.
3. Improved Talent Acquisition and Retention: By offering this attractive work arrangement, [Company Name] expects to reduce turnover costs and attract high-caliber talent, further enhancing our competitive edge.
4. Enhanced Employee Well-being: Reduced burnout and improved work-life balance are expected to result in higher quality work, increased creativity, and better customer service.

"We understand that this change may seem unconventional, but we are committed to staying ahead of the curve in our industry," added [CEO Name]. "Our financial projections indicate that this move will positively impact our profitability within the first year of implementation, with even greater returns expected in subsequent years."[Company Name] will provide regular updates to shareholders and the public on the progress and impact of this initiative. The company remains dedicated to delivering value to its shareholders while fostering a progressive and productive work environment.

For more information, please contact:
[PR Contact Name]
[PR Contact Email]
[PR Contact Phone]
About [Company Name]:
[Brief company description]

Request for Review

Still here? I have a small favor to ask. Would you consider sharing your thoughts at https://www.amazon.com/gp/product/B0DK41RWZS? Your review can make a world of difference to both the author and future readers. It only takes two minutes, but it means a lot. Simply head over to https://www.amazon.com/gp/product/B0DK41RWZS or point your cell phone camera at the QR Code (below), tap the link and let others know what you think. Whoa, did you notice the rhyme? Thank you. Thank you.

Your opinion matters. Your words could inspire someone else to discover their next favorite read. Thanks in advance.

Section Quiz

Life is not a spectator sport. Here's a quiz.

1. **What was the reported increase in productivity at Microsoft Japan after implementing a 4-day workweek?**

a) 20%
b) 30%
c) 40%
d) 50%

2. **According to the University of Cambridge study, by what percentage did burnout reduce with a 4-day workweek?**

a) 39%
b) 65%
c) 71%
d) 82%

3. **What percentage of companies reported reduced staff turnover after implementing a 4-day workweek?**

a) 32%
b) 50%
c) 79%

d) 91%

4. **How much could a company potentially save on variable overhead expenses with a 4-day workweek?**
a) Up to 10%
b) Up to 15%
c) Up to 20%
d) Up to 25%

5. **What percentage of employees consider flexible work schedules an extremely important benefit?**
a) 50%
b) 65%
c) 79%
d) 91%

6. **According to a UK study, by what percentage could a 4-day workweek potentially shrink the country's carbon footprint?**
a) 15.7%
b) 21.3%
c) 22.2%
d) 24.1%

7. **How many full-time employees does the company in the scenario have?**
a) 1,000
b) 5,000
c) 10,000
d) 15,000

8. **What is the annual revenue of the company mentioned in the scenario?**
a) $100 million
b) $200 million
c) $300 million
d) $500 million

9. **According to the text, how much can employee turnover cost a company?**
a) Up to one times an employee's salary
b) Up to two times an employee's salary
c) Up to three times an employee's salary
d) Up to four times an employee's salary

10. **In the press release, what is the proposed work schedule for employees under the new policy?**
a) Four 8-hour days per week
b) Four 9-hour days per week
c) Four 10-hour days per week
d) Five 6-hour days per week

Answers

1. c) 40%
2. c) 71%
3. b) 50%
4. c) Up to 20%
5. c) 79%
6. b) 22.2%
7. c) 10,000
8. a) $100 million
9. b) Up to two times an employee's salary
10. a) Four 8-hour days per week

Case Study: Net Neutrality

Background

How about another highly controversial item these days, "net neutrality." If you're asking, "What is that?" you're in the right place.

I am copying the thread from Perplexity with no content editing.[xxii] Sometimes the formatting is off. I am still keeping it as-is. Watch the follow-ups. My prompts are in quotes. Perplexity added graphics to its answers. I removed some graphics because they had too much verbiage or were blurry.

There will be 10 review questions at the end. Enjoy.

Tell me About Net Neutrality

"What is net neutrality?"

Net neutrality is the principle that Internet service providers (ISPs) must treat all Internet communications equally, without discriminating or charging differently based on content, user, platform, application, type of equipment, or method of communication. This concept ensures an open and fair Internet for everyone, regardless of the device, application, or platform used and the content consumed.

Key Aspects of Net Neutrality

Equal Access: Net neutrality requires ISPs to provide the same level of data access and speed to all traffic. They cannot block, throttle, or create "fast lanes" for specific websites or services.

No Discrimination: Under net neutrality, ISPs are prohibited from favoring certain content providers or charging them for preferential treatment. This helps maintain a level playing field for businesses and promotes innovation.

Freedom of Information: Net neutrality supports the free exchange of ideas and knowledge by preventing ISPs from controlling what users can access or read online.

Historical Context

The term "net neutrality" was coined by Columbia University law professor Tim Wu in 2003. It has since become a central issue in Internet policy discussions:

- In 2015, the Obama administration enacted net neutrality rules in the United States.
- These rules were repealed by the Trump administration in 2017.
- In 2024, the Biden administration, through the Federal Communications Commission (FCC), reinstated net neutrality rules.
- However, in August 2024, these rules were blocked by courts.

Ongoing Debate

The net neutrality debate remains contentious, with supporters and opponents presenting various arguments:

Supporters argue that net neutrality:

- Promotes innovation and fair competition
- Protects freedom of speech and information access
- Prevents ISPs from creating unfair "fast lanes" for content providers who can afford to pay more

Opponents claim that net neutrality:

- Reduces investment in broadband infrastructure
- Imposes unnecessary regulations
- Hinders ISPs' ability to manage network traffic efficiently

As of January 2025, the future of net neutrality regulations in the United States remains uncertain, with ongoing legal challenges and potential changes in future administrations.

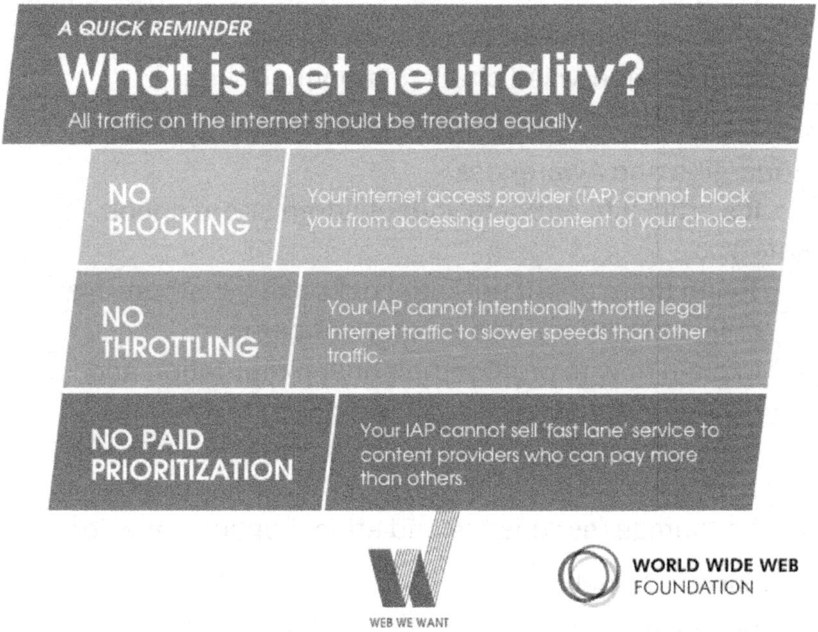

--

"I do not work for an ISP. How does net neutrality help me keep my grade school children safe online?"

Net neutrality itself is not directly related to keeping children safe online. However, it does provide an important foundation for equal access to online educational resources, which can indirectly support a safer online experience for children. That said, there are several ways you can help keep your grade school children safe online:

Parental Controls and Monitoring

Many Internet Service Providers (ISPs) offer parental control features that allow you to:

- Create multiple user profiles for different family members

- Set up filtering options to block specific sites or keyword searches https://www.ncta.com/whats-new/establishing-a-safer-internet-experience-for-children
- Implement time limits on screen time
- Enable supervised chats to monitor online conversations.

Education and Awareness

It's crucial to educate your children about online safety:

- Teach them about potential risks like cyberbullying, online predators, and privacy concerns
- Explain how to protect their online reputation and digital footprint https://www.ncta.com/whats-new/empowering-families-create-safer-internet-environments-for-kids
- Encourage responsible and ethical online behavior

Family Guidelines

Establish clear rules for internet use:

- Never post or trade personal pictures online https://kidshealth.org/en/parents/net-safety.html
- Don't reveal personal information like address, phone number, or school name
- Use only screen names and don't share passwords (except with parents) https://kidshealth.org/en/parents/net-safety.html
- Never agree to meet someone in person without parental approval and supervision
- Always tell a trusted adult about any scary or hurtful online communication https://kidshealth.org/en/parents/net-safety.html

Parental Supervision

Stay involved in your children's online activities:

- Keep computers in common areas where you can monitor their use https://kidshealth.org/en/parents/net-safety.html
- Spend time online together to teach appropriate behavior
- Bookmark favorite sites for easy access
- Take your child seriously if they report uncomfortable online exchanges https://kidshealth.org/en/parents/net-safety.html

Cybersecurity Measures

Protect your children from online threats:

- Use security software packages, including antivirus and firewall solutions
- Teach children about the risks of downloading unknown files or apps https://www.ncta.com/whats-new/empowering-families-create-safer-internet-environments-for-kids
- By implementing these strategies, you can create a safer online environment for your grade school children while still allowing them to benefit from the educational opportunities the internet provides.

"Help me explain net neutrality to my kids and write your response at a 6th grade reading level."

Net neutrality is like a fair playground for the internet. Imagine the internet as a big playground where everyone can play and share their ideas. Net neutrality makes sure that everyone gets to play equally, no matter who they are or what games they like.

What Net Neutrality Does

Treats Everyone the Same: With net neutrality, all websites and apps are treated equally. It's like everyone getting a turn on the swings, no matter if they're big or small.

No Special Treatment: Internet companies can't make some websites faster or slower than others. It's like not letting some kids cut in line just because they're friends with the teacher.

Keeps Things Fair: Net neutrality helps new ideas and websites grow. It's like giving new kids a chance to join in games, even if they're not as popular yet.

Why It's Important

Net neutrality helps keep the internet fun and fair for everyone. Here's why it matters:

- You can visit any website you want without it being blocked or slowed down.

- New and cool websites have a chance to become popular.

- You don't have to pay extra to use your favorite apps or watch videos.

An Easy Way to Understand It

Think of the internet like getting a milkshake:

1. Imagine you're drinking a free milkshake, but you need to buy a straw.
2. The straw seller starts pinching your straw, making it hard to drink.
3. They say they'll only stop if the milkshake shop pays them more money.
4. But you already paid for the straw, and it's not fair to make it harder for you to enjoy your milkshake.

This is kind of like what could happen without net neutrality. Some websites might be harder to use unless they pay extra, which isn't fair to you or the websites you like. Remember, net neutrality is all about keeping the internet open and fair for everyone, just like a playground where all kids get to play together equally.

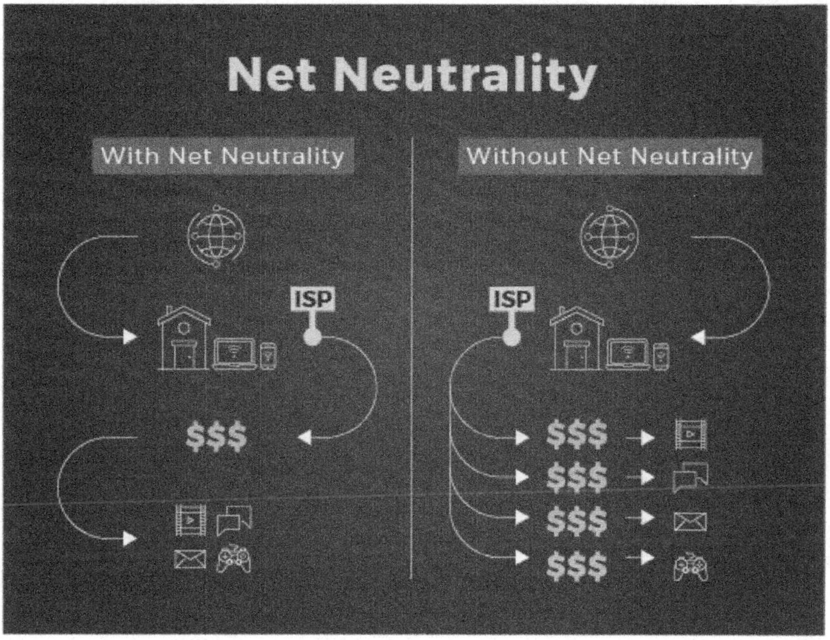

--

"I work at a bank. Help me explain net neutrality to the Vice Presidents of my bank. They all have MBAs."

Net neutrality is a critical principle that ensures fair and equal treatment of all internet traffic, regardless of its source or destination. For financial institutions like our bank, understanding net neutrality is essential due to its potential impact on our operations, customer interactions, and competitive landscape.

Key Aspects of Net Neutrality

Equal Access: Net neutrality mandates that Internet Service Providers (ISPs) treat all data on the internet equally, without discriminating or charging differently based on user, content, website, platform, application, type of equipment, or method of communication.

Regulatory Framework: The Federal Communications Commission (FCC) classifies broadband service as a "telecommunications" service under Title II, giving it oversight over internet service outages, national security threats in broadband networks, and consumer protection.

Implications for Banking

Digital Services: Net neutrality ensures that our online banking platforms and mobile apps can be accessed by customers without preferential treatment or throttling by ISPs https://www.techtarget.com/searchnetworking/definition/Net-neutrality.

Competition: It provides a level playing field for financial institutions of all sizes, preventing larger competitors from gaining unfair advantages through paid prioritization agreements with ISPs. **Innovation**: Net neutrality fosters an environment conducive to fintech innovation, allowing new financial products and services to reach customers without facing artificial barriers.

Business Considerations

Cost Management: Without net neutrality, banks might face additional costs for ensuring optimal delivery of their digital services, potentially impacting operational expenses.

Customer Experience: Consistent internet speeds and access are crucial for maintaining high-quality customer experiences across our digital platforms.

Data Security: Net neutrality rules can enhance the FCC's authority to implement cybersecurity standards and protect consumer data, which is paramount in banking.

Potential Risks of Repealing Net Neutrality

Increased Costs: ISPs could potentially charge higher fees for prioritized access to financial services websites and apps.

Competitive Disadvantage: Smaller banks and fintech startups might struggle to compete if larger institutions can afford preferential treatment from ISPs.

Regulatory Uncertainty: Changes in net neutrality policies could lead to shifts in how financial data is transmitted and accessed, requiring adaptations in compliance strategies.

Strategic Considerations

1. **Digital Infrastructure**: Evaluate our current digital service delivery and assess potential vulnerabilities in a non-neutral internet scenario.
2. **Partnerships**: Consider strategic alliances with ISPs or tech companies to ensure optimal service delivery in various regulatory environments.
3. **Innovation Focus**: Invest in cutting-edge technologies that can adapt to potential changes in internet traffic management.
4. **Regulatory Engagement**: Actively participate in policy discussions to ensure the banking sector's interests are represented in net neutrality debates.

Understanding net neutrality is crucial for our bank's strategic planning, as it directly impacts our ability to serve customers efficiently and compete effectively in an increasingly digital financial landscape.

--

"I am in favor of net neutrality. Write me three articles on net neutrality for my blog. Make sure they are between 200 and 400 words."

Here are three articles on net neutrality for your blog, each between 200-400 words:

The Importance of Net Neutrality for a Free and Open Internet

Net neutrality is the principle that all internet traffic should be treated equally by Internet Service Providers (ISPs). This concept is crucial for maintaining a free and open internet where everyone has equal access to information and services.

Under net neutrality rules, ISPs are prohibited from blocking, throttling, or creating "fast lanes" for specific websites or services. This ensures that small startups have the same opportunity to reach users as large, established companies. It prevents ISPs from favoring their own content or services over those of competitors.Net neutrality also protects free speech online. Without these protections, ISPs could potentially slow down or block access to websites with opposing political views or content they disagree with. This could lead to a fragmented internet where access to information is determined by your service provider.

For consumers, net neutrality means you can access any legal content without additional fees or reduced speeds. It prevents scenarios where you might have to pay extra to stream videos or use certain applications.

While opponents argue that net neutrality stifles innovation and investment, proponents contend that it actually fosters innovation by providing a level playing field for new ideas and services to flourish.

As we continue to rely more heavily on the internet for work, education, and daily life, preserving net neutrality becomes increasingly important. It's not just about faster Netflix streaming – it's about ensuring equal access to information and opportunities in our digital world.

Net Neutrality: A Cornerstone of Digital Democracy

In our increasingly connected world, net neutrality stands as a fundamental principle of digital democracy. It ensures that the internet remains a platform for free expression, innovation, and equal opportunity.Net neutrality rules prevent Internet Service Providers (ISPs) from discriminating against certain types of content or favoring specific websites or services. This means that whether you're accessing a small personal blog or a major news website, your ISP must treat that traffic equally.

One of the key benefits of net neutrality is its role in fostering competition and innovation. Without these protections, large companies could potentially pay ISPs for faster access to consumers, creating an unfair advantage over smaller competitors or startups. This "pay-to-play" model could stifle the next big idea before it even has a chance to take off.

For consumers, net neutrality means freedom of choice. It ensures that you can access the content and services you want without interference from your ISP. This is particularly important as more of our daily activities – from work and education to entertainment and social interaction – move online.

Critics argue that net neutrality regulations are unnecessary and could hamper investment in network infrastructure. However, proponents point out that the internet thrived under net neutrality rules, and that these protections are essential for preserving the open nature of the internet.

As we navigate the complexities of our digital age, maintaining net neutrality is crucial for ensuring that the internet remains a tool for empowerment and equality, rather than becoming a system that reinforces existing power structures and inequalities.

The Global Perspective on Net Neutrality

While much of the net neutrality debate has centered on the United States, it's important to recognize that this is a global issue with far-reaching implications for internet users worldwide.

Different countries have taken varying approaches to net neutrality. The European Union, for instance, has strong net neutrality protections in place. In contrast, some countries have weaker or no net neutrality rules, potentially allowing ISPs more control over internet traffic.

The global nature of the internet means that decisions about net neutrality in one country can have ripple effects across borders. For example, if a major country weakens its net neutrality protections, it could influence global internet norms and potentially lead to a more fragmented internet.

In developing countries, net neutrality is particularly crucial. Without these protections, ISPs could potentially create tiered internet services, where only those who can afford to pay more have access to the full internet. This could exacerbate existing digital divides and hinder economic development.Net neutrality also has implications for freedom of expression on a global scale. In countries with strong net neutrality rules, it's harder for governments or corporations to censor or control access to information. This makes the internet a powerful tool for democracy and human rights.

As we look to the future, maintaining and strengthening net neutrality protections around the world will be essential for preserving the internet as a global public resource. It's not just about technical rules – it's about ensuring that the internet continues to be a platform for global communication, innovation, and opportunity for all.

--

"I am in favor of net neutrality. Write me three articles on net neutrality that I will post on LinkedIn. Make sure they are between 200 and 400 words."

Here are three articles on net neutrality suitable for posting on LinkedIn, each between 200-400 words:

Net Neutrality: Safeguarding Innovation in the Digital Economy

In today's rapidly evolving digital landscape, net neutrality stands as a crucial pillar supporting innovation and fair competition. As professionals, we must understand its significance for our industries and the broader economy.Net neutrality ensures that Internet Service Providers (ISPs) treat all online content equally, without favoring or blocking particular websites or services. This level playing field is vital for startups and small businesses, allowing them to compete with established players based on the merit of their ideas rather than their ability to pay for preferential treatment.

Consider the impact on emerging technologies like AI, IoT, and cloud computing. Without net neutrality, ISPs could potentially slow down or block services that compete with their own offerings, stifling innovation and limiting consumer choice. This could have far-reaching consequences for industries relying on these technologies for growth and efficiency.

For professionals in tech, finance, healthcare, and other sectors, net neutrality ensures equal access to online resources, platforms, and potential customers. It allows for the free flow of ideas and information, fostering collaboration and innovation across industries.

Moreover, net neutrality supports remote work capabilities, online learning platforms, and digital transformation initiatives – all critical components of the modern workplace.

As business leaders and professionals, we should advocate for strong net neutrality protections. It's not just about preserving an open internet; it's about ensuring a fair, competitive, and innovative digital economy that benefits businesses of all sizes and drives economic growth.

The Business Case for Net Neutrality

As professionals and business leaders, understanding the implications of net neutrality on our operations and growth strategies is crucial. Net neutrality, the principle that all internet traffic should be treated equally, has significant impacts on how we conduct business in the digital age.

Without net neutrality, Internet Service Providers (ISPs) could potentially create "fast lanes" for companies willing to pay more, leaving smaller competitors at a disadvantage. This pay-to-play model could dramatically alter the competitive landscape across industries.

Consider the impact on customer experience. In a world without net neutrality, the quality and speed of your company's website or app could depend on deals struck with ISPs rather than the merit of your service. This could lead to customer frustration and loss of business to competitors who can afford preferential treatment.

For startups and small businesses, net neutrality is particularly crucial. It allows them to reach customers on an equal footing with larger, more established companies. This fosters innovation and ensures that the best ideas, not the biggest budgets, drive success.Net neutrality also supports the growing trend of remote work and digital collaboration. Equal access to high-speed internet is essential for teams spread across different locations, ensuring seamless communication and productivity.

From a cost perspective, the absence of net neutrality could lead to increased operational expenses. Businesses might need to pay for prioritized access to ensure their services remain competitive, potentially impacting profit margins.

As business professionals, we should recognize that net neutrality is not just a technical issue, but a fundamental principle that supports fair competition, innovation, and economic growth in our increasingly digital world.

Net Neutrality and the Future of Work

As professionals, we're witnessing a transformation in how and where work gets done. Remote work, digital nomadism, and the gig economy are reshaping our professional landscape. At the heart of these changes lies the principle of net neutrality, which ensures equal access to the internet for all.Net neutrality prevents Internet Service Providers (ISPs) from discriminating against certain types of content or favoring specific websites or services. This is crucial for the future of work in several ways:

1. Equal Opportunities: Net neutrality ensures that freelancers and remote workers can access job platforms, collaboration tools, and clients without facing artificial barriers or increased costs.
2. Innovation in Work Tools: It allows new productivity and collaboration tools to emerge and compete fairly, driving innovation in how we work.
3. Skill Development: Equal access to online learning platforms enables professionals to upskill and reskill, adapting to changing job market demands.
4. Diverse Workforce: By ensuring equal internet access, net neutrality supports a more diverse and inclusive workforce, allowing talent from various geographical locations to contribute.
5. Small Business Growth: It levels the playing field for small businesses and startups, allowing them to reach customers and partners without being outspent by larger competitors.

Without net neutrality, we risk a future where the quality of our work tools and opportunities depends on our ability to pay for prioritized internet access. This could exacerbate existing inequalities in the job market and hinder economic mobility.

As professionals, we should advocate for strong net neutrality protections. It's not just about preserving an open internet; it's about ensuring a fair, innovative, and inclusive future of work that benefits all professionals, regardless of their location or resources.

--

"I am in favor of net neutrality. Write me three posts on net neutrality for my Instagram. Suggest pictures or short videos to go with the verbiage."

Here are three Instagram posts on net neutrality, each with a suggested image or video:

1. Post:

Net neutrality = internet freedom! 🌐 ✊
It means your ISP can't slow down Netflix just because they prefer Hulu. It keeps the internet fair for everyone, from big companies to small startups. Without it, your favorite apps might load slower or cost more. Let's keep the net neutral!
#NetNeutrality #InternetFreedom #DigitalRights

Image suggestion: A simple graphic showing two lanes of internet traffic. One lane is labeled "With Net Neutrality" and shows all data moving at the same speed. The other is labeled "Without Net Neutrality" and shows some data zooming ahead while other data crawls along.

2. Post:

Did you know? Net neutrality protects your right to access ANY legal content online without interference.
📱 💻
It's like having an all-access pass to the internet highway. No toll booths, no speed bumps for certain websites. Keep the internet open and equal for all!
#NetNeutrality #OpenInternet #DigitalEquality

Video suggestion: A short animation showing a person browsing different websites (represented by icons) at equal speeds. Then show what it might look like without net neutrality - some icons loading quickly, others buffering or blocked.

3. Post:

Net neutrality isn't just about streaming your favorite shows. It's about:

- Protecting free speech online 🗣
- Fostering innovation and competition 💡
- Ensuring equal access to information 📚
- Preventing ISPs from controlling what you see 👀

Stand up for a free and open internet!
#NetNeutrality #InternetFreedom #DigitalRights

Image suggestion: A collage of four images representing the four points mentioned: a megaphone for free speech, a lightbulb for innovation, an open book for information access, and an eye for preventing ISP control.

Section Quiz

Life is not a spectator sport. Here's a quiz.

1. **What is the main principle of net neutrality?**
a) Faster internet for everyone
b) Equal treatment of all internet traffic
c) Free internet access for all
d) Government control of the internet

2. **Which of the following is NOT a key aspect of net neutrality?**
a) No discrimination of content
b) Equal access for all users
c) Freedom of information
d) Guaranteed high-speed internet for all users

3. **Who coined the term "net neutrality"?**
a) Bill Gates
b) Tim Berners-Lee
c) Tim Wu
d) Mark Zuckerberg

4. **How does net neutrality affect competition among businesses?**
a) It eliminates all competition
b) It gives large companies an advantage
c) It creates a level playing field for businesses of all sizes
d) It only benefits tech companies

5. **Which of the following is a potential risk of repealing net neutrality?**
a) Improved internet speeds for all users
b) Increased costs for preferential treatment of content
c) More government control over the internet
d) Reduced use of social media platforms

6. **How might net neutrality affect innovation?**
a) It stifles innovation by over-regulating the internet
b) It has no effect on innovation
c) It promotes innovation by providing equal opportunities for new ideas

d) It only encourages innovation in large tech companies

7. **What analogy was used to explain net neutrality to children?**
a) A highway with different speed limits
b) A library with restricted access to certain books
c) A playground where everyone gets to play equally
d) A store with different prices for the same product

8. **How does net neutrality impact freedom of speech online?**
a) It has no effect on free speech
b) It limits free speech by regulating content
c) It protects free speech by preventing ISPs from blocking or slowing down content
d) It only protects political speech

9. **In the context of banking, how does net neutrality affect digital services?**
a) It makes online banking impossible
b) It ensures bank websites can be accessed without preferential treatment
c) It gives larger banks an advantage in digital services
d) It has no impact on banking services

10. **How does net neutrality relate to the future of work?**
a) It has no impact on work environments
b) It only affects tech industry jobs
c) It supports remote work and digital collaboration by ensuring equal internet access
d) It makes traditional office work more important

Answers

1. b) Equal treatment of all internet traffic
2. d) Guaranteed high-speed internet for all users
3. c) Tim Wu

4. c) It creates a level playing field for businesses of all sizes
5. b) Increased costs for preferential treatment of content
6. c) It promotes innovation by providing equal opportunities for new ideas
7. c) A playground where everyone gets to play equally
8. c) It protects free speech by preventing ISPs from blocking or slowing down content
9. b) It ensures bank websites can be accessed without preferential treatment
10. c) It supports remote work and digital collaboration by ensuring equal internet access

Case Study: Cyber Insurance

Background

Here's the last highly controversial item, cyber insurance. I am copying the thread from Perplexity with no content editing.[xxiii] Sometimes the formatting is off. I am still keeping it as-is. Watch the follow-ups. My prompts are in quotes. I asked Perplexity for answers in Spanish and English. Watch for those at the end of this section. Perplexity added graphics to its answers. I removed some graphics because they had too much verbiage or were blurry.

There will be 10 review questions at the end. Enjoy.

Should We Get Cyber Insurance?

"What is cyber insurance?"

Cyber insurance, also known as cyber liability insurance or cyber risk insurance, is a specialized insurance product designed to protect businesses from financial losses and liabilities associated with Internet-based risks and information technology infrastructure. This type of insurance has become increasingly important as the digital landscape continues to evolve and cyber threats become more prevalent.

Key Features of Cyber Insurance

Coverage Areas:

Cyber insurance typically covers a range of potential losses and expenses, including:

- Data breaches and theft
- Cyber attacks, including ransomware and hacking
- Business interruption due to cyber events

- Extortion attempts
- Reputational damage
- Legal fees and regulatory fines

Types of Coverage:

Cyber insurance policies often include two main components:

1. First-party coverage: This protects the insured business from direct losses and expenses resulting from a cyber incident.
2. Third-party coverage: This provides protection against claims made by customers, partners, or other external parties affected by the cyber event.

Importance of Cyber Insurance

The need for cyber insurance has grown significantly due to several factors:

- Increasing frequency and sophistication of cyber attacks
- Rising costs associated with data breaches (average cost of $4.35 million per breach) https://www.ibm.com/think/topics/cyber-insurance
- Potential for significant financial and reputational damage
- Limited coverage for cyber risks in traditional insurance policies

What Cyber Insurance Typically Covers

Cyber insurance policies can vary, but common areas of coverage include:

- Forensic investigations
- Data recovery and system repair
- Notification costs for affected individuals
- Credit monitoring services

- Public relations expenses
- Legal defense and settlement costs
- Regulatory fines and penalties
- Business interruption losses
- Cyber extortion payments

Limitations and Considerations

It's important to note that cyber insurance is not a substitute for robust cybersecurity measures. Most policies exclude coverage for preventable issues or those caused by human error or negligence. Businesses should view cyber insurance as a complement to their overall risk management strategy, not a replacement for strong security practices.

In conclusion, cyber insurance provides a vital safety net for businesses operating in the digital age, helping to mitigate the financial impact of cyber incidents and supporting recovery efforts in the aftermath of an attack.

--

"My boss asked me for the pros and cons of cyber insurance. He wants me to present this at a manager's meeting. Give me the pros and cons of cyber insurance in a format I can easily convert into Microsoft PowerPoint format."

Pros of Cyber Insurance

- Financial Protection
- Covers costs associated with data breaches and cyber attacks
- Includes recovery expenses, legal fees, and business interruption losses
- Specialized Expertise
- Provides access to cybersecurity specialists for incident response

https://www.morganfranklin.com/insights/understanding-the-pros-and-cons-of-cybersecurity-insurance/

- Offers support services for crisis management and public relations
- Regulatory Compliance
- Assists with meeting data privacy law requirements
- Covers costs of customer notifications and regulatory penalties
- Business Continuity
- Reimburses lost income due to business interruptions
- Helps safeguard short and long-term business future
- Improved Security Posture
- Encourages adoption of best practices in cybersecurity
- May include regular security audits and risk assessments

Cons of Cyber Insurance

- Cost Considerations
- Can be expensive, especially for high-risk businesses
- Pricing may be inconsistent among insurers
- Coverage Limitations
- May have exclusions or limits on certain types of losses
- Might not cover intellectual property loss or reputational damage
- Complexity
- Policies can be complex and difficult to understand
- May require detailed knowledge of internal protection levels
- False Sense of Security

- Not a substitute for robust cybersecurity measures
- May lead to complacency in implementing security practices
- Potential Restrictions
- May impose business-impacting restrictions based on policy specifications
- Insurer might dictate specific remediation steps

--

"Some of the managers have only high school diplomas. Write me the pros and cons at a 6th grade reading level."

Pros of Cyber Insurance

- Money Protection
- Helps pay for costs if your computer systems are attacked
- Covers things like fixing damaged systems and lost business money
- Expert Help
- Gives you access to smart people who can help if you're hacked
- These experts can help fix problems and talk to customers
- Legal Support
- Helps pay for lawyers if someone sues you because of a cyber attack
- Can cover fines you might have to pay
- Keeps Business Going
- Pays you back for money lost when your business has to stop because of an attack
- Helps your business survive after a big computer problem

Cons of Cyber Insurance

- It's Expensive
- Can cost a lot of money, especially for small businesses
- Prices keep going up as more cyber attacks happen
- Doesn't Cover Everything
- Might not pay for all types of computer attacks
- Some policies have limits on how much they'll pay
- Can Be Confusing
- Policies can be hard to understand
- You might need help figuring out what's covered and what's not
- False Safety Feeling
- Having insurance doesn't mean you're safe from attacks
- You still need to protect your computers and train your workers

Remember, cyber insurance is just one part of keeping your business safe online. It's important to also have good computer security practices.

La Respuesta de la Perplejidad en Español

"Some managers speak Spanish better than they speak English. Give me the same answer in Spanish."

Ventajas del Seguro Cibernético

- Protección Financiera
- Ayuda a pagar los costos si tus sistemas informáticos son atacados
- Cubre cosas como reparar sistemas dañados y dinero perdido del negocio
- Ayuda Experta
- Te da acceso a personas inteligentes que pueden ayudar si te hackean

- Estos expertos pueden ayudar a solucionar problemas y hablar con los clientes
- Apoyo Legal
- Ayuda a pagar abogados si alguien te demanda por un ataque cibernético
- Puede cubrir multas que tengas que pagar
- Mantiene el Negocio Funcionando
- Te devuelve el dinero perdido cuando tu negocio tiene que parar por un ataque
- Ayuda a que tu negocio sobreviva después de un gran problema informático

Desventajas del Seguro Cibernético

- Es Caro
- Puede costar mucho dinero, especialmente para negocios pequeños
- Los precios siguen subiendo a medida que ocurren más ataques cibernéticos
- No Cubre Todo
- Podría no pagar por todos los tipos de ataques informáticos
- Algunas pólizas tienen límites en cuánto pagarán
- Puede Ser Confuso
- Las pólizas pueden ser difíciles de entender
- Podrías necesitar ayuda para entender qué está cubierto y qué no
- Falsa Sensación de Seguridad
- Tener seguro no significa que estés a salvo de ataques
- Todavía necesitas proteger tus computadoras y entrenar a tus trabajadores

Recuerda, el seguro cibernético es solo una parte de mantener tu negocio seguro en línea. Es importante también tener buenas prácticas de seguridad informática.

Perplexitys Antwort auf Deutsch

"Some managers speak German better than they speak English. Give me the same answer in German."

Vorteile der Cyber-Versicherung

- Finanzieller Schutz
- Hilft bei der Bezahlung von Kosten, wenn Ihre Computersysteme angegriffen werden
- Deckt Dinge wie die Reparatur beschädigter Systeme und verlorenes Geschäftsgeld ab
- Expertenunterstützung
- Gibt Ihnen Zugang zu klugen Leuten, die helfen können, wenn Sie gehackt werden
- Diese Experten können bei der Problemlösung und Kundenkommunikation helfen
- Rechtliche Unterstützung
- Hilft bei der Bezahlung von Anwälten, wenn Sie wegen eines Cyberangriffs verklagt werden
- Kann Strafen abdecken, die Sie möglicherweise zahlen müssen
- Hält das Geschäft am Laufen
- Erstattet Ihnen Geld, das verloren geht, wenn Ihr Geschäft wegen eines Angriffs stoppen muss
- Hilft Ihrem Unternehmen, nach einem großen Computerproblem zu überleben

Nachteile der Cyber-Versicherung

- Es ist teuer
- Kann viel Geld kosten, besonders für kleine Unternehmen
- Die Preise steigen weiter, da mehr Cyberangriffe passieren
- Deckt nicht alles ab
- Könnte nicht für alle Arten von Computerangriffen zahlen

- Einige Policen haben Grenzen, wie viel sie zahlen werden
- Kann verwirrend sein
- Policen können schwer zu verstehen sein
- Sie könnten Hilfe brauchen, um zu verstehen, was abgedeckt ist und was nicht
- Falsches Sicherheitsgefühl
- Eine Versicherung zu haben bedeutet nicht, dass Sie vor Angriffen sicher sind
- Sie müssen immer noch Ihre Computer schützen und Ihre Mitarbeiter schulen

Denken Sie daran, dass die Cyber-Versicherung nur ein Teil davon ist, Ihr Geschäft online sicher zu halten. Es ist auch wichtig, gute Computersicherheitspraktiken zu haben.

Section Quiz

Life is not a spectator sport. Here's a quiz.

1. **What is the primary purpose of cyber insurance?**

a) To replace cybersecurity measures
b) To provide financial protection against cyber risks

c) To prevent cyber attacks
d) To train employees in cybersecurity

2. **Which of the following is typically covered by cyber insurance?**
a) Regular software updates
b) Employee salaries
c) Data breach costs
d) Office rent

3. **What type of expertise does cyber insurance often provide access to?**
a) Marketing specialists
b) Cybersecurity experts
c) Financial advisors
d) Human resources professionals

4. **Which of the following is a potential drawback of cyber insurance?**
a) It's too cheap
b) It provides too much coverage
c) It can be expensive
d) It's easy to understand

5. **How does cyber insurance help with business continuity?**
a) By providing new computers
b) By reimbursing lost income due to business interruptions
c) By hiring new employees
d) By relocating the business

6. **What kind of legal support does cyber insurance typically offer?**
a) Help with divorce proceedings
b) Support for criminal cases
c) Assistance with property disputes
d) Coverage for legal fees related to cyber incidents

7. **What is a potential negative consequence of having cyber insurance?**
a) Increased cyber attacks
b) Lower profits
c) False sense of security
d) Higher employee turnover

8. **What aspect of cyber insurance policies can be challenging for businesses?**
a) They're too short
b) They can be complex and difficult to understand
c) They change daily
d) They're written in code

9. **In addition to having cyber insurance, what else is important for a business?**
a) Having good computer security practices
b) Ignoring cybersecurity altogether
c) Relying solely on insurance
d) Keeping all data public

10. **What type of costs might cyber insurance cover in the event of a data breach?**
a) Only hardware replacement
b) Only software updates
c) Only legal fees
d) Recovery expenses, legal fees, and business interruption losses

Answers

1. b) To provide financial protection against cyber risks
2. c) Data breach costs
3. b) Cybersecurity experts
4. c) It can be expensive
5. b) By reimbursing lost income due to business interruptions

6. d) Coverage for legal fees related to cyber incidents
7. c) False sense of security
8. b) They can be complex and difficult to understand
9. a) Having good computer security practices
10. d) Recovery expenses, legal fees, and business interruption losses

AI-Driven Dynamic Pricing

Corporate Greed or Digital Revolution?

Price Changes Specific to You

Imagine this: you walk into your local Kroger grocery store, heading straight for the sports drink aisle to grab some bottles for the upcoming week. But wait! The price tag on your favorite sports drink suddenly changes right before your eyes. Sounds like science fiction, right? Well, it's not. It is dynamic pricing. It is already happening!

Kroger, one of the biggest players in the grocery game with over 2,700 stores, is using something called AI-driven dynamic pricing. It's like having a super smart computer decide how much things should cost. It can change those prices faster than you can say "I better put on my glasses!"

Here's how dynamic pricing works:

- Cameras in the store spot you when you enter the store.
- The AI remembers what you usually buy.
- AI looks at things like the time of day, how busy the store is and even the weather.
- Then, faster than you can blink an eye, it changes the prices on the shelves.

Kroger says this is all to make shopping better for us. Is it corporate greed or a digital revolution?

The AI is Keeping Track of What?

Let us break down what this AI is recording:

- The day and time you shop.
- Your age (or at least a guess at it).
- What you buy most often.
- How much you usually spend.

Is AI-Driven Dynamic Pricing Fair?

Now, I know what you're thinking: "Mr. Sugartooth, how is this fair?" Well, that's the million-dollar question, isn't it? Let's go back to the sports drink example. You're a big fan of a certain brand of sports drink. You buy it every week. The AI notices this and thinks, "Hey, this customer really loves this drink. I bet they'd pay a little more for it." Suddenly, the price goes up by a few cents.

Let us revisit the question, “Is AI-driven dynamic pricing fair?”

Kroger Says Yes

This is where things get a bit sticky, like pine tar on a baseball bat. Kroger says this system helps them offer better deals and keep prices low overall. Here’s how Kroger justifies its dynamic pricing strategy, according to Perplexity.ai:

Lowering prices over time

Kroger consistently states that their business model and dynamic pricing strategy aim to lower prices for customers in the long run. They argue that this approach allows them to offer better deals to consumers over time.

Improving customer experiences

The company claims that their dynamic pricing strategy is designed to enhance the overall shopping experience for customers.

Increasing revenue for reinvestment

Kroger asserts that by lowering prices, they attract more customers, which leads to increased revenue. They then claim to reinvest this additional revenue back into the business, potentially for further price reductions or improvements.

Targeted price reductions

Kroger states that any testing of electronic shelf labels is intended to lower prices more effectively for customers "where it matters most". This suggests they aim to offer better deals on products that are most important to their customers.

Denying price gouging accusations

In response to concerns about potential price gouging, Kroger has explicitly stated that suggestions of using dynamic pricing to exploit consumers are "not true".

While Kroger presents these justifications, some critics and lawmakers remain skeptical about the potential negative impacts of dynamic pricing on consumers. They argue that the system could be used to increase prices during peak demand periods or exploit consumer data for profit. Before advancing to the next page, let me ask, “What do you think about Kroger’s argument?”

Consumer Advocates Say No

Opponents think it's just a way for big companies to make more money off their most loyal customers. Here's the argument against AI-driven dynamic pricing, also from Perplexity.AI:

Economic Inequality and High Prices

Kroger's new pricing system might make things harder for people who don't have a lot of money. This system could try to get the most money from each customer. This could hurt people who are already struggling to buy food. Some people in the government are worried about this too.

Two senators, Elizabeth Warren and Bob Casey, are looking into how Kroger sets its prices. They're worried that if more stores use digital price tags, big grocery stores might charge more to make more money. They think this could lead to grocery prices changing like airplane tickets do, with higher prices when stores are busy or for things people really need.

Privacy Worries

Another big problem with Kroger's pricing system is that it might invade people's privacy. The system uses computers to set prices, which makes people wonder how much of their personal information is being used. Some worry that the system might track what people buy and use that to set prices.

Unclear Pricing

It's hard to understand why prices keep changing with this new system. This can make people not trust the store. Customers might feel like the store is tricking them or raising prices for no good reason. Sometimes perception speaks louder than reality.

Unfair Pricing Concerns

Some people are worried that this pricing system might not be fair to everyone. If the system uses information about individual shoppers or neighborhoods to set prices, it might charge more to certain groups of people or in certain areas. This brings up questions about whether it's fair and if everyone has the same chance to buy things.

How It Affects Shoppers and Store Loyalty

Some people think changing prices all the time might cause problems with how people shop. One person on the internet said that rich people might stop shopping at Kroger, while poor people might come expecting low prices. This could hurt Kroger's business in the long run.

Problems with Making It Work

It is not easy to change prices quickly in real stores. Unlike online shops or airlines, grocery stores need to change price tags by hand. This can cost a lot of money and take a lot of time. People wonder if this system can really work well in a grocery store.

Claims of Greedy Business

Some groups that protect shoppers think Kroger is being too greedy. They say that changing prices like this, along with trying to join with other big stores, shows that Kroger cares more about making money than helping customers.

Worries About Controlling Prices

People are worried that Kroger might use this system to control prices unfairly. By changing prices quickly based on what people want or how much of something they have, Kroger could make things seem scarce or raise prices when people need things most, like before storms or holidays.

What Do You Think?

Kroger says their new pricing system can lead to lower prices and happier customers. Many people think it could hurt shoppers, especially those who don't have much money. As this new way of pricing keeps changing, groups that protect shoppers and people who make rules will watch closely to see how it affects stores and shoppers.

Armed with this latest information, let me ask, "What do you think about Kroger's argument now?"

Dynamic Pricing in Other Industries Too

Other Big Names Do It

Fasten your seatbelts. Retail is not the only industry with AI-driven dynamic pricing. Here are some others. I'm sure you will recognize some of the big names.

Hospitality and Travel

Airbnb's dynamic pricing tool, "Smart Pricing," uses machine learning to suggest optimal prices to hosts based on location, seasonality and local events.

Marriott International uses a system that analyzes factors such as local events, weather and historical booking patterns to adjust room rates in real-time.

Expedia employs AI to analyze vast amounts of data across multiple travel services, adjusting prices for flights, hotels and rental cars based on demand and availability.

Delta Airlines uses dynamic pricing to offer personalized fares based on a traveler's history, loyalty status and current demand.

Uber's surge pricing model is a well-known example of dynamic pricing, where ride prices increase during periods of high demand to incentivize more drivers to become available.

Retail

Target uses AI to analyze competitor pricing, inventory levels and demand patterns to adjust prices across its vast product range.

Best Buy implements dynamic pricing particularly for electronics, where prices can fluctuate rapidly based on new product releases and competitor actions.

Home Depot employs dynamic pricing for seasonal items, adjusting prices based on local demand and inventory levels.

Macy's uses a system that adjusts prices based on inventory levels, sales performance and even weather patterns that might affect clothing demand.

Kroger made the top 5 list. Imagine that. We've already discussed what Kroger does.

E-commerce

Amazon uses sophisticated algorithms to adjust prices on millions of items multiple times per day. Their system considers factors such as competitor pricing, supply and demand and even the time of day to optimize prices.

Walmart employs similar tactics, using AI to analyze vast amounts of data to set competitive prices across its online platform.

eBay's approach focuses on auction-style listings, where AI helps determine optimal starting prices and reserve prices based on historical data and current market trends.

Etsy caters to handmade and vintage items. Etsy uses AI to suggest pricing to sellers and dynamically adjust featured item prices based on demand and search trends.

Wayfair uses dynamic pricing to manage its vast inventory of home goods, adjusting prices based on popularity, seasonality and competitor offerings.

Entertainment and Media

Disney+ employs similar tactics, using AI to determine optimal pricing for different regions and content offerings.

Netflix uses AI to analyze viewing habits and adjust subscription prices in different markets. While they don't change prices frequently for individual users, they use dynamic pricing strategies when entering new markets or introducing new plan tiers.

Ticketmaster uses dynamic pricing for event tickets, adjusting prices based on demand, which can sometimes lead to controversial price spikes for popular events.

StubHub, a ticket resale platform, uses AI to suggest optimal pricing to sellers and adjust fees based on market demand.

Spotify, while primarily subscription-based, uses AI-driven dynamic pricing for its advertising rates, adjusting prices based on listener demographics and ad engagement rates.

Manufacturing and Industrial

Yes, big names in the B2B (business to business) world do dynamic pricing too. Maybe nobody notices, or nobody has bothered to say anything.

Caterpillar uses dynamic pricing for its heavy machinery and parts, adjusting prices based on global demand, currency fluctuations and regional economic conditions.

General Electric uses dynamic pricing to adjust prices for its industrial equipment based on factors such as customer segment, order volume and market conditions.

John Deere implements dynamic pricing for its agricultural equipment, considering factors such as crop prices, seasonal demand and local economic conditions.

Boeing uses AI-driven pricing models for its commercial aircraft, adjusting prices based on long-term market forecasts, fuel prices and airline profitability projections.

Siemens employs AI to optimize pricing for its diverse range of products, considering factors like raw material costs, demand forecasts and competitor pricing.

In all these industries, AI-driven dynamic pricing allows companies to respond quickly to market changes, optimize revenue and offer more personalized pricing to customers.

AI-driven dynamic pricing changes prices automatically. We're not always talking about a rainy day that cancels your double-header. For the B2B industries, we are talking about a heavy rainy season that causes floods and forces farmers to abandon their crops. The crops that survive the bad weather end up costing you and me more.

Agriculture

What is more important than food? Weather impacts crop production and of course, crop losses. The American Farm Bureau monitors crop production. AI-driven dynamic pricing models digest this information and adjust prices accordingly.

Do you remember paying more for corn in the summer of 2022? Severe weather increased crop losses. What do you suppose AI-driven dynamic pricing did to prices for those who like corn? It was worse in 2024. Severe weather increased crop losses. Almost a $20 billion increase in two years. Farmers had to recoup their losses somehow.

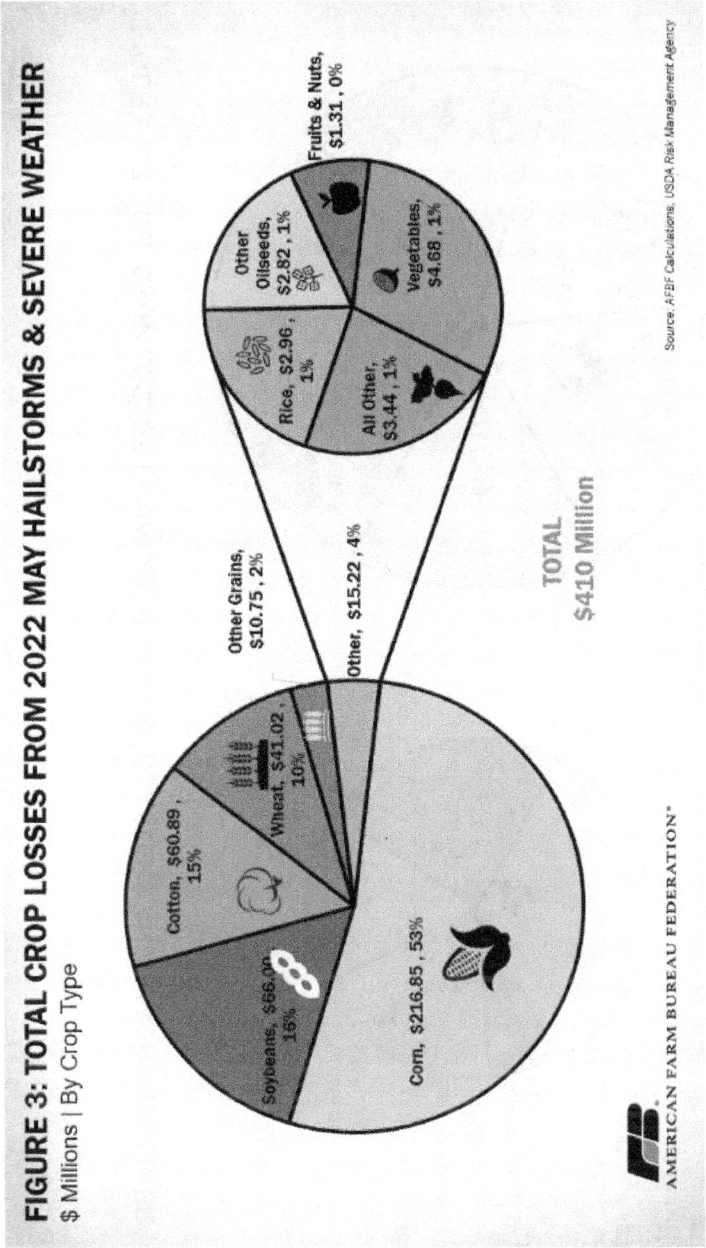
FIGURE 3: TOTAL CROP LOSSES FROM 2022 MAY HAILSTORMS & SEVERE WEATHER
$ Millions | By Crop Type
Soybeans, $66.0
16%
Cotton, $60.89, 15%
Wheat, $41.02, 10%
Corn, $216.85, 53%
Other Grains, $10.75, 2%
Other, $15.22, 4%
Rice, $2.96, 1%
Other Oilseeds, $2.82, 1%
Fruits & Nuts, $1.31, 0%
All Other, $3.44, 1%
Vegetables, $4.68, 1%
TOTAL
$410 Million
AMERICAN FARM BUREAU FEDERATION
Source: AFBF Calculations, USDA Risk Management Agency

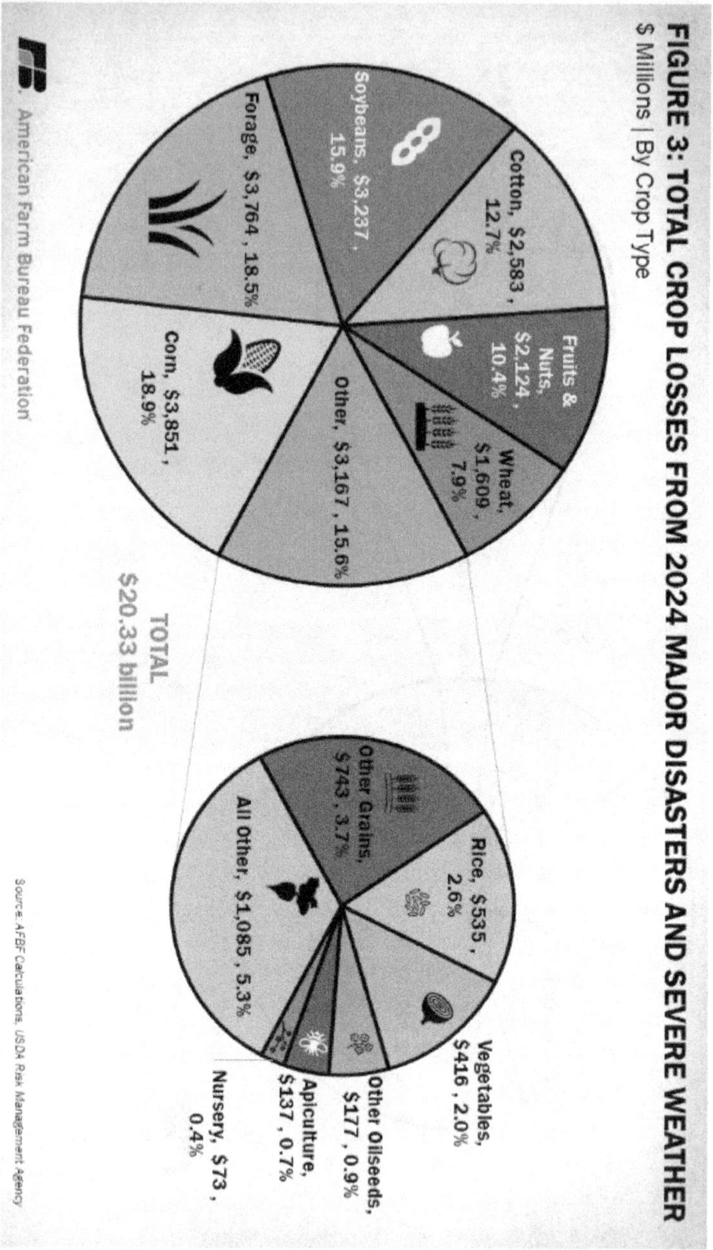
FIGURE 3: TOTAL CROP LOSSES FROM 2024 MAJOR DISASTERS AND SEVERE WEATHER
$ Millions | By Crop Type
Cotton, $2,583, 12.7%
Fruits & Nuts, $2,124, 10.4%
Wheat, $1,609, 7.9%
Other, $3,167, 15.6%
Corn, $3,851, 18.9%
Forage, $3,764, 18.5%
Soybeans, $3,237, 15.9%
TOTAL
$20.33 billion
Rice, $535, 2.6%
Vegetables, $416, 2.0%
Other Oilseeds, $177, 0.9%
Apiculture, $137, 0.7%
Nursery, $73, 0.4%
All Other, $1,085, 5.3%
Other Grains, $743, 3.7%
American Farm Bureau Federation
Source: AFBF Calculations, USDA Risk Management Agency

What You Will See

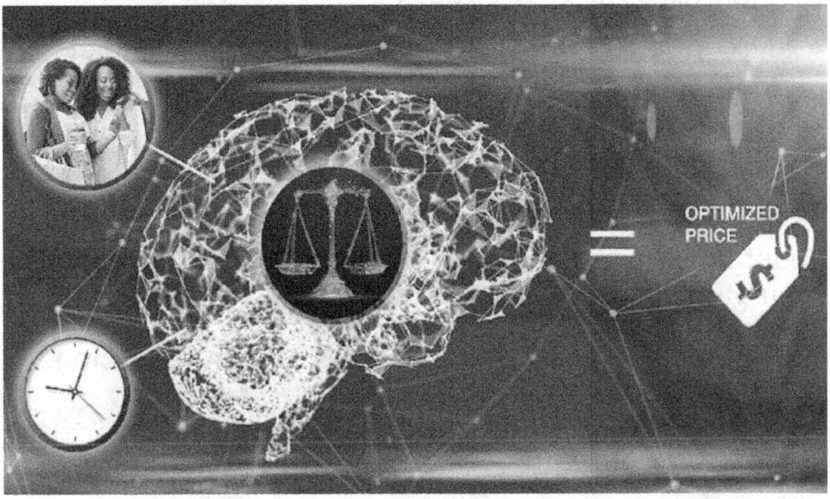

Let's talk about some of the tricky tactics AI-driven dynamic pricing can do:

- It might raise prices on things you buy often, thinking you'll pay more for your favorites.
- It could lower prices on items you've never tried, hoping to get you hooked.
- The AI might send you special coupons for things it thinks you'll like.
- Your shopping habits could be sold to other companies who want to sell you their products.

You are the product. Maybe I should say, "Your information is the product." If sellers know which brand of sports drink you prefer and how much you usually pay, they'll use that to their advantage.

What Can You Do About It?

You might not be able to change the whole system, but you can shop smarter. Here are some tips to protect your wallet:

- Keep your eye on the prices.
- Mix up your shopping routine so the AI can't predict what you'll do.
- Compare prices at different stores.
- Use coupons and sales to your advantage.

Remember, knowledge is power. The more you know about how these systems work, the better you can play their game!

How Delta Airlines Uses AI-Driven Dynamic Pricing

I'm sure you noticed Delta Airlines back in the Hospitality and Travel list. They use AI-driven dynamic pricing too. My last professional trip involved flying from Sacramento to Atlanta, and back. Let's use that route as an example:

Airlines use dynamic pricing to adjust ticket prices based on various factors, including a passenger's shopping habits. An airline tracks your browsing history and notices you've searched for flights from Sacramento to Atlanta multiple times. The airline's algorithm interprets this as high interest in the route and potential willingness to pay.

For a Sacramento to Atlanta route on Delta, the award prices can vary dramatically. In one instance, the same route might cost:

- 17,000 miles at a lower-demand time, OR
- 36,500 miles during peak travel periods.

The pricing directly correlates with cash fares. The same route might cost:

- $198 at a lower-demand time, OR
- $399 during peak travel periods.

Airlines consider multiple elements when dynamically pricing tickets:

- Day of the week.
- Time of search.
- Seat availability.
- Seasonal demand.
- Your previous booking patterns.

Practical Tips To combat dynamic pricing:

- Be flexible with travel dates.
- Use the airline's price calendar.
- Book during less popular travel times.
- Compare prices across different days.

The goal is to find the most cost-effective option by understanding how airlines adjust prices in real-time based on complex algorithms and market conditions.

Anonymous Shopping

Remember when I wrote, "The airline's algorithm interprets this as high interest in the route and potential willingness to pay" in the last section? That same strategy applies in other industries too.

Suppose you're shopping but not ready to make a purchase decision today. Do not let the sellers know how much interest you have. Shop anonymously.

- For Delta, do not sign in to the app, or website, when shopping. Sign in only when ready to make a purchase decision.
- Which other apps and websites do you regularly use for online shopping? Let me guess. Amazon. eBay. Home Depot. Staples. Starbucks. Target. Walmart. They do not need to know when you're merely shopping. Verify you are **not** signed in when merely shopping.
- Shop anonymously until you're ready for a purchase decision. Then and only then do you sign in and make the purchase. The retailer will **not** know how much time you've spent shopping or how interested (or desperate) you are to make a purchase decision.

"But Mr. Sugartooth, I am a frequent shopper and have rewards," you may say. Good. After you sign in, you'll see whatever loyalty rewards and preferential pricing are waiting for you.

Get Ahead of the Curve

How closely are you watching prices? Are price tags on products and store shelves static, or do they have electronic displays that update before your eyes? Some people don't even look for the price. They want a dozen oranges. They grab a dozen oranges.

Will your favorite retailers start using AI-driven dynamic pricing, or will Elizabeth Warren stop them? Will there be rules to make sure it is fair for everyone, even those who are desperate?

Section Summary

So, what do you think about all this? Forgive the Seller? The seller is taking a big gamble here. Selling you something at a big discount is better than not selling you that item at all. The seller might even be selling you something at a price lower than they paid. They are willing to take a loss today in the hopes that you'll buy more tomorrow. On the flip side, they could argue increasing the price you pay is necessary to help them pay their bills. After all, your favorite retailers are running businesses too.

Corporate Greed? The first version of *AI-Driven Dynamic Pricing* (where I pulled most of this section's content from) had the subtitle *A Grand Slam for Corporate Greed.* Do you think that was fair, too aggressive, or too polite? Send your thoughts to mark@cybersafetynet.net.

Section Quiz

Life is not a spectator sport. Here's a quiz.

1. **What is AI-driven dynamic pricing primarily used for in Kroger stores?**
a) To keep all prices constant
b) To change prices based on customer behavior and other factors
c) To always offer the lowest prices
d) To eliminate the need for cashiers

2. **Which of the following is NOT mentioned as something the AI system tracks?**
a) The time of day you shop
b) Your age (or an estimate of it)
c) What you buy most often
d) Your shoe size

3. **What is suggested as a way for shoppers to outsmart the AI pricing system?**
a) Always shop at the same time
b) Buy only the most expensive items
c) Mix up your shopping routine
d) Avoid using coupons

4. **Which company is prominently featured in the text as using AI-driven dynamic pricing?**
a) Walmart
b) Target
c) Kroger
d) Amazon

5. **What is one of the justifications Kroger provides for using dynamic pricing?**
a) To increase profits
b) To offer better deals and keep prices low overall
c) To confuse customers
d) To eliminate competition

6. **Who is mentioned as a critic of AI-driven dynamic pricing in the document?**
a) Senator Elizabeth Warren
b) Mr. Sugartooth
c) The CEO of Kroger
d) Mark Zuckerberg

7. **In which industry, besides retail, is Airbnb's "Smart Pricing" mentioned as an example of dynamic pricing?**
a) Manufacturing
b) Entertainment
c) Agriculture
d) Hospitality and Travel

8. **What is suggested as a potential negative consequence of AI-driven dynamic pricing for loyal customers?**
a) They might receive more discounts
b) They might pay higher prices for favorite items
c) They might be banned from the store
d) They might receive worse customer service

9. **According to the text, what can you do to stop online retailers from tracking your interest?**
a) Shop outside of regular business hours
b) Browse websites in Incognito mode
c) Access apps and websites anonymously
d) Do all your shopping in person

10. **What does the document suggest about the future of AI-driven dynamic pricing?**
a) It will be banned soon
b) It will only be used in grocery stores
c) Its future use and regulation are uncertain
d) It will be mandatory for all businesses

Answers

1. b) To change prices based on customer behavior and other factors
2. d) Your shoe size
3. c) Mix up your shopping routine
4. c) Kroger
5. b) To offer better deals and keep prices low overall
6. a) Senator Elizabeth Warren
7. d) Hospitality and Travel
8. b) They might pay higher prices for favorite items
9. c) Access apps and websites anonymously
10. c) Its future use and regulation are uncertain

Powering AI

Power Hungry AI

How power hungry is artificial intelligence? AI is so power hungry that Microsoft and other AI companies are turning to nuclear power to meet the demand. Microsoft is revitalizing an old nuclear power plant, with a bad reputation, to generate energy for Microsoft's AI Data Centers.

Three Mile Island is Now Called TMI Energy Park

First off, let's talk about Three Mile Island. It is old and was taken out of service. It is famous. I'll talk about why... later. It's getting a major makeover and is now called TMI Energy Park. This old power plant is set to become a hub of clean, nuclear energy, and Microsoft is driving this initiative. Why? Because they're all about powering their tech with clean, reliable energy.

Microsoft's Energy Adventure

Microsoft has set some pretty ambitious goals. They want to be carbon neutral this year. That means they're working hard to reduce their carbon footprint and use more renewable energy sources. By investing in TMI Energy Park, they're not just buying energy; they're buying into a cleaner, greener future.

Microsoft's Background

Bill Gates and Paul Allen founded Microsoft in 1975. They built DOS (Disk Operating System) and by 1981, their operating system became the default operating system of personal computers. By the late 1980s, Microsoft was the largest company in the personal computer software development space. In 2014, Microsoft promoted Satya Nadella to become the new CEO. He made cloud computing a high priority and is spearheading the company's surge into AI. Good things have happened on his watch. Microsoft's stock value has increased with a reported 27% annual growth rate since he became CEO. They reported record revenue of $211 billion in fiscal year 2023. They also had a $3.394 trillion market capitalization[xxiv] as of July 2024. Do they just spend their profits on the world's largest office party? No. They reinvest. One place they reinvest their profits is in AI.

Why Nuclear?

Nuclear energy is like having a power source that never takes a day off. It's highly reliable and can provide consistent energy 24/7. This is called baseload power, meaning it's always there when you need it. Unlike solar or wind, which depend on the weather, nuclear keeps on chugging along no matter what. Microsoft needs this kind of steady power to keep their data centers running smoothly. These centers are the backbone of everything from Xbox gaming to cloud computing. Without reliable energy, AI and the Internet in general could become unreliable.

Like Microsoft's servers, several items in your home need steady power. If your home loses steady power, you incur a blackout. A blackout occurs when the energy supply is not strong enough to meet the demand. This happens in summer when it is super-hot outside and people run the A/C units to cool their homes. It also happens at Christmastime, as neighbors turn on Christmas decorations without thinking about how much energy each miniature light bulb consumes. If a blackout occurs, several devices that usually work all the time stop working.

- Computers. They rely on steady power to keep them running. When they do not have steady power, computers turn off on their own. No advance warning.
- Appliances. In addition to the obvious desktop computer, your home has other items that are computers on the inside. This includes washing machines. Last year, I replaced the system board in my 22-year-old washing machine. The system board looked like a computer's system board.
- HVAC (Heating, Ventilation Air Conditioning). Remember a few lines back when I wrote about blackouts? Well, your furnace and air conditioner (or heat pump) rely on constant energy. You may go a whole month without your air conditioner turning on, but when you want it on, you really want it. If your power utility cannot meet the demand from your air conditioner, your neighbors' air conditioners and other items that are drawing energy, your neighborhood experiences a blackout.

When the energy demand is greater than what the transformer can handle, BOOM! Out go the lights. This usually happens in the summer. When it is hot and your neighbors all run their air conditioners, the unmeetable demand causes a blackout. People lose their cool. Get it? When air conditioners do not get enough energy, people lose their cool.

Why Pennsylvania?

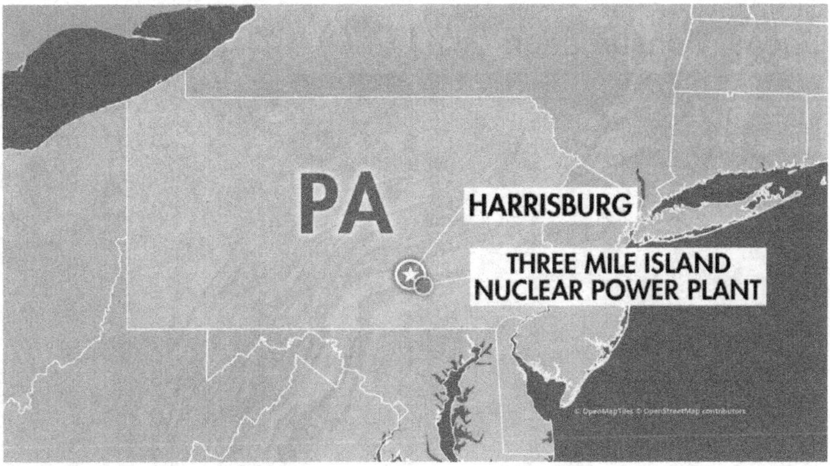

So why did Microsoft choose Pennsylvania? Pennsylvania isn't just known for Taylor Swift, cheesesteaks and the Liberty Bell. It is also home to TMI Energy Park. This location is part of the PJM Interconnection, a regional transmission organization that helps distribute energy efficiently across multiple states. By choosing Pennsylvania, Microsoft can tap into this network and power its data centers across the Northeast more effectively. Plus, the state has a supportive regulatory environment for nuclear energy, making it an ideal spot for Microsoft's clean energy plans.

Energy or Electricity?

To help keep things clear, I am using the term *energy* and not *electricity* whenever possible.

When?

Turning on a nuclear reactor is easier said than done. If testing goes well, and if the U.S. Nuclear Regulatory Commission says yes, TMI Energy Park unit 1 will resume operating in 2028.

AI vs. Google Search: The Energy Race

Energy Consumption

Now, let's talk about energy consumption. Microsoft uses AI servers that consume a lot more energy than traditional search servers like Google's. Here's how they stack up:

AI Servers: These are like the monster trucks of computing, powerful but energy hungry. They use up to 30 times more energy than Google Search servers and generate a lot of heat. For example, you visit https://perplexity.ai and prompt "How much water should a 50-year-old man drink each day and the man works outside and the high temperature this week will exceed 100 degrees Fahrenheit and what recommendations do you have for appropriate clothing and respond at a 6th grade reading level." We jump for joy because the answer is complete and helpful. The technologies that gave you this wonderful answer did much more work than a Google search requires. That is why they use up to **30 times more energy** than Google Search servers and generate a lot of heat.

Google Search Servers: These are more like Toyota Priuses, still powerful but consume less energy. For example, you visit https://google.com and search for "How much water should I drink a day?" You get a list of 10 web pages. We were overjoyed at one time. These days, blah.

AI Prompt	**Google Search**
"How much water should a 50-year-old man drink each day and the man works outside and the high temperature this week will exceed 100 degrees Fahrenheit and what recommendations do you have for appropriate clothing and respond at a 6th grade reading level."	"How much water should I drink every day?"
Result: Actionable content.	Result: List of web pages.
Uses 30 units of energy.	Uses 1 unit of energy.

Nuclear is Consistent and Reliable

Nuclear is highly consistent and reliable. That's why it supports baseline demand. No matter the weather, environment or time of day, there will always be baseline demand for energy. This baseline demand comes from HVAC systems, water heaters, refrigerators, freezers, lights, media equipment and of course, computers. Let's also remember the computers you access when you visit websites, perform Google searches and perform AI research are working 24 hours a day, seven days a week.

"But Mr. Sugartooth, what about wind and solar. They provide clean energy too." I am glad you asked.

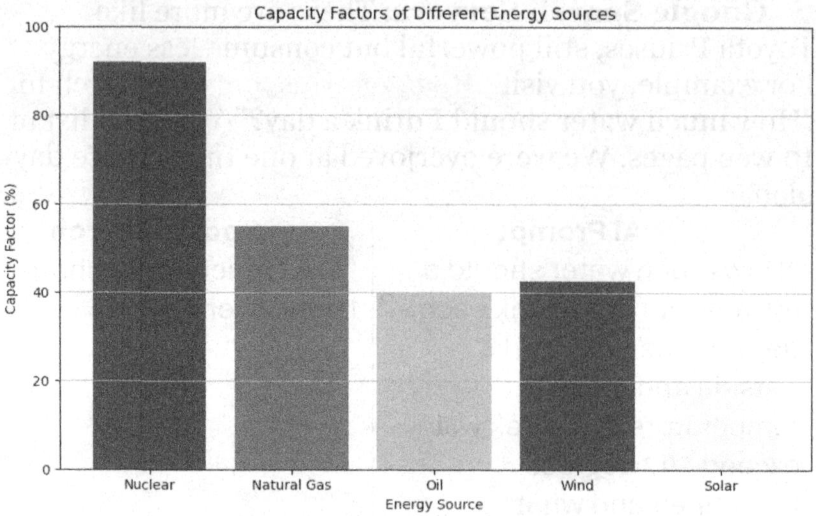

Nuclear is most reliable. Weather impacts wind and solar production. Wind has obvious Upsides and Downsides. I've seen fields of windmills lined up nice and neat to capture wind in their blades, spin turbines and produce energy. We call these wind farms.

Upside of Generating Energy from Wind

Wind energy is a clean and renewable power source. It doesn't produce greenhouse gases, which helps reduce pollution and combat climate change. Wind turbines also have low operating costs once installed because they don't need fuel. Wind farms can provide economic benefits to local communities because they create jobs and generate lease payments to landowners.

Downside of Generating Energy from Wind

The main downside is unpredictability. Wind doesn't always blow consistently, making it an unreliable sole energy source. Wind turbines can also impact wildlife. Sometimes birds and bats fly into windmills. Wind farms require specific locations with strong winds. We cannot just build wind farms anywhere.

Upside of Generating Energy from Solar

Solar energy is a clean and renewable power source. It doesn't produce pollution or greenhouse gases, helping to fight climate change. Solar panels can be installed on rooftops, reducing the need for large power plants and lowering energy bills. They have low maintenance costs and can provide energy independence.

Downside of Generating Energy from Solar

The main downside is that solar power depends on sunlight, so it doesn't work well at night or on cloudy days. Solar panels also require a lot of space and can be expensive to install. Additionally, their efficiency can decrease in areas with less sunlight.

Which Energy Source is Best?

The election of 2024 has come and gone. Good. One item of contention that arises every election cycle is energy production. Which forms of energy deserve investment? Which ones don't? Let me summarize these options in a table:

Not 100% Great	**Not 100% Awful**
Coal	Coal
Gas	Gas
Geothermal	Geothermal
Hydroelectric	Hydroelectric
Nuclear	Nuclear
Oil	Oil
Solar	Solar
Wind	Wind

Every option has its strengths and weaknesses. We need it all. Climate change is making the planet hotter. AI adoption is creating demand for more AI data centers.

Must Keep Their Cool

Server Rooms Need Cooling

Because AI servers generate so much heat, they need extra cooling systems to keep them from overheating. Imagine wearing a sweater in summer. That's how these servers feel without proper cooling.

AI servers reside in data centers. These are rooms that hold servers and other network hardware that keeps everything working. These server rooms also have advanced air management strategies that must keep the room comfortable for the servers **and** for people.

How They Keep Their Cool

"How cool?" you may ask. I deployed a workstation last week. The manufacturer's spec sheet said the maximum operating temperature was 95 degrees Fahrenheit. I've seen some server ratings with a maximum operating temperature of 113 degrees Fahrenheit.,

Suppose you had 2,000 computers in one room. They were all running at the same time. You signed a *Service Level Agreement* to a) provide sufficient electricity and b)

provide sufficient colling to keep these computers from melting. That is a tall order.

When air flow and cooling are correctly planned, air movement is highly tweaked. Cold air enters the room and has nowhere to go accept through the racks where it picks up the heat from servers and other IT equipment. Air exits the IT equipment, enters the hot aisle, rises and gets pulled back to the HVAC equipment. Hot air is sucked out of the room, only to be replaced by cold air. The cycle repeats.

The next page has an illustration showing airflow in a server room. Study how cold air enters, flows through servers, draws heat, leaves the servers, leaves the room and returns to the chiller.

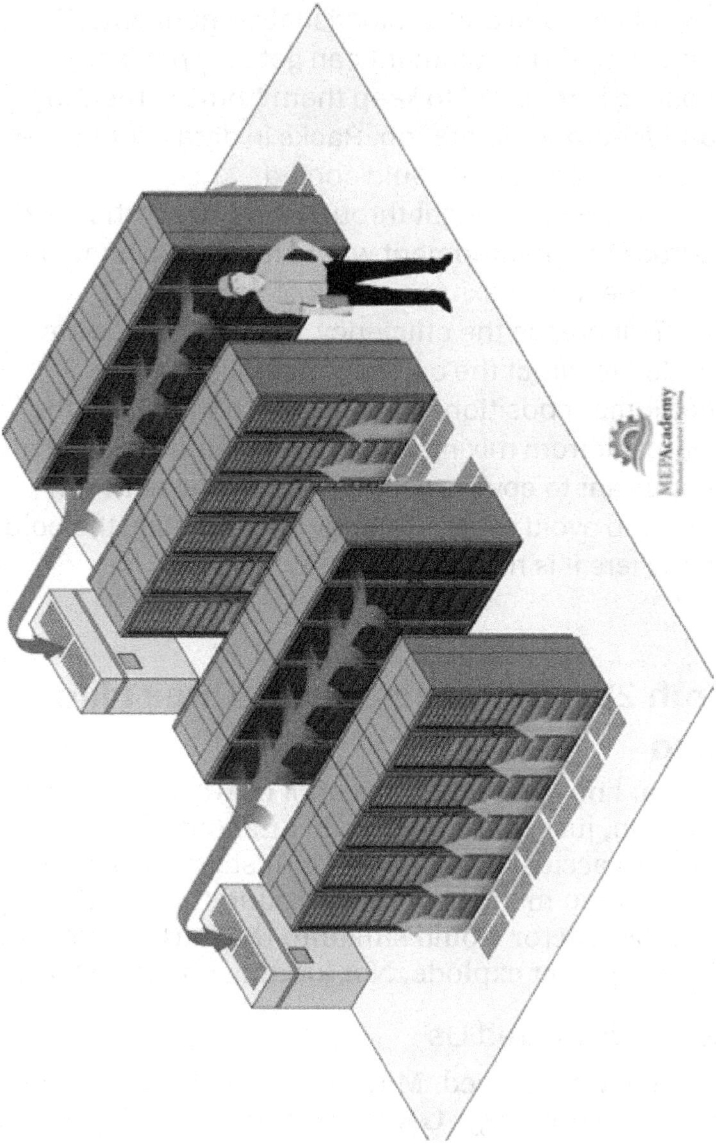

The folks at MEP Academy[xxv] further tell us:

> All the IT equipment sits on shelves arranged vertically in a rack. The standard rack height is 7 feet (2.1m). These racks lined up together in neat rows in data centers. The racks house and protect data center equipment such as servers, routers,

switches, hubs and audio/visual components. The data center IT equipment can get very hot, so cooling is required to keep them from overheating and for proper operation. Racks in data centers are either air-cooled or liquid cooled.

Cold air is brought through the front of the rack, across the IT equipment where it picks up heat, and then the hot air exits the back of the rack.

To increase the efficiency, blanking plates are added to direct the cold air optimally over the IT equipment positioned in the rack, and to keep the warm air from mixing with the cold entering air. It's important to cover openings within and between racks to avoid wasting energy and directing the cold air where it is needed.

March 28, 1979, Accident at Three Mile Island

If you know anybody at school or work who has a bad reputation, just wait until you hear about Three Mile Island. An accident and series of mistakes at Three Mile Island in 1979 made headlines around the world. People feared the reactor would sink into the Earth's core and make our planet explode. Not kidding about this one.

Jane Fonda Scared Us

People were scared. Maybe because they knew little about nuclear energy, **OR** because they saw Jane Fonda, Michael Douglas and Jack Lemmon star in *The China Syndrome*, which was released 12 days before the Three Mile Island accident.

Many governmental agencies documented the errors at Three Mile Island. You could spend an entire weekend reading all of them, only to learn they have different viewpoints and all are right.

What Perplexity Reports

The Three Mile Island accident on March 28, 1979, involved several mistakes made by people. These mistakes made the problem worse. Here's a simple explanation of what went wrong, according to Perplexity.ai:

Operator Mistakes

The people running the power plant made some big errors:

- They didn't understand what was really happening. They thought it was a different kind of problem than it actually was.
- They misread an important valve. They thought it was closed when it was actually open. This was because they didn't have a good way to see if it was really closed or not.

- They turned off emergency cooling water because they were worried about overfilling a part of the reactor. This was the wrong thing to do.
- They didn't notice that some important valves were closed when they should have been open. This stopped emergency water from flowing where it was needed.
- They didn't look at all the available information. There was a temperature gauge that could have helped them figure out the problem, but they didn't check it.

Training and Rules Problems

The workers weren't prepared well enough for this kind of emergency:

- They hadn't been trained properly to handle this specific situation.
- The rules and instructions they had weren't good enough to guide them through...

Communication Issues

There were also problems with how information was shared:

- The plant managers didn't tell state officials quickly enough about steam coming out of the plant. This made people not trust them.
- There was a lot of confusion when people were talking on the phone about what was happening. This led to misunderstandings about how serious the situation was.

All these human errors, along with equipment problems, caused the accident to get worse. It showed how important it is to have good training, clear communication and strong safety rules in nuclear power plants.

What happened is well-documented. Regardless, we need energy. The Three Mile Island incident's human errors are well-documented. We have learned. We now know more. We have more oversight.

Section Summary

AI needs power. The forecast calls for even more power demand tomorrow. I prompted Perplexity, "What is the forecasted demand for energy for AI-servers over the next 10 years?" Fasten your seat belts:

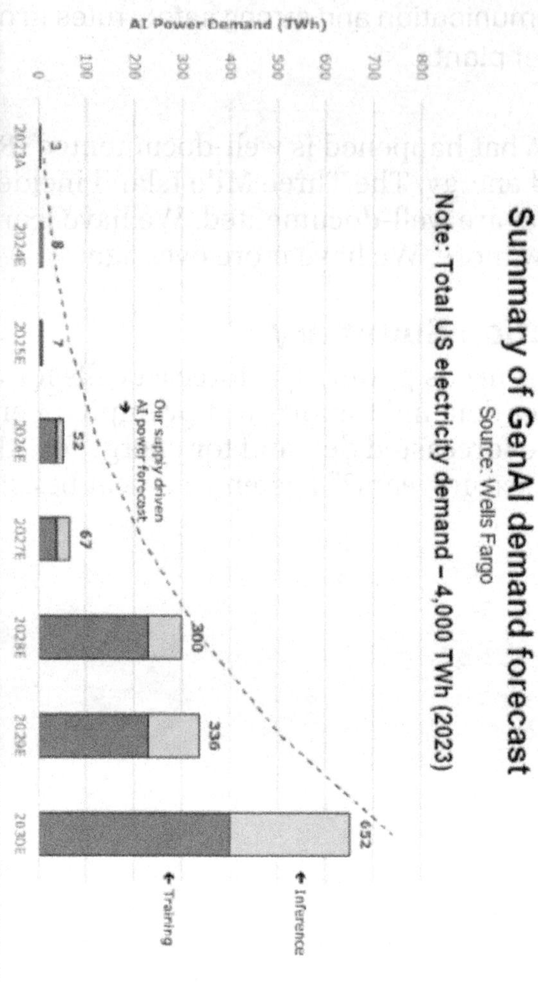

	2023A	2024E	2025E	2026E	2027E	2028E	2029E	2030E
(+) Training power demand (TWh)	3	8	7	40	40	229	229	402
(+) Inference power demand (TWh)	0	0	0	12	27	70	107	250
Demand driven AI power forecast (TWh)	**3**	**8**	**7**	**52**	**67**	**300**	**336**	**652**
Supply driven AI power forecast (TWh)	9	29	65	125	217	341	508	739

Source: Wells Fargo Securities, LLC estimates

Microsoft and other data center providers have their hands full. Can they contact other old/decommissioned nuclear power plants and ask them to get back up and running? Can utilities build new power plants fast enough to meet this demand? Is our nation comfortable enough with nuclear power to allow more power plants, or will the lack of energy restrict AI's growth? Only time will tell.

Three days after I initially published *Powering AI*, Google announced they would power new data centers through nuclear power. Those plants will be smaller and of course newer than Three Mile Island. Powering AI is a race to meet demand.

What are your thoughts? Send your thoughts to mark@cybersafetynet.net.

Flipping a wall switch and turning on a light creates an energy demand. Visiting Google.com and searching for "wall-mounted dimmer switches" creates additional demand. Going to Perplexity and prompting "Show me how a wall-mounted dimmer works, compared to a toggle switch **and** write at a 6th grade reading level" creates even more energy demand. We need to figure out how to meet energy demand. We're already at a deficit.

Our nation has brought only two nuclear power plants online since 2014 (when Satya Nadella became Microsoft CEO). Both are at the Vogtle Electric Generating Plant, approximately 188 miles southeast of Atlanta. I think we have some catch up to do.

Section Quiz

Life is not a spectator sport. Here's a quiz.

1. **What is the new name of Three Mile Island?**

a) TMI Energy Park
b) Three Mile Renaissance
c) Crane Clean Energy Center

d) Energy Island

2. **Why did Microsoft choose to buy energy in Pennsylvania?**
a) Lower taxes
b) Access to nuclear energy
c) Proximity to headquarters
d) Abundant solar resources

3. **Which energy source has the highest capacity factor according to the chart?**
a) Natural Gas
b) Wind
c) Nuclear
d) Solar

4. **What is Microsoft's goal for its energy consumption by 2025?**
a) Double its usage
b) Achieve carbon neutrality
c) Use only fossil fuels
d) Reduce costs by 50%

5. **Why is nuclear energy considered reliable?**
a) It depends on the weather conditions
b) It has a high-capacity factor and provides baseload power
c) It is the cheapest energy source
d) It requires minimal infrastructure

6. **Which of these energy sources is most dependent on sunshine?**
a) Oil
b) Solar
c) Natural Gas
d) Nuclear

7. **What significant event occurred at Three Mile Island in 1979?**
a) A major solar breakthrough
b) A nuclear incident
c) The opening of a wind farm
d) An oil spill

8. **How does AI server energy consumption compare to Google Search servers?**
a) AI servers use less energy
b) They use about the same amount of energy
c) AI servers use significantly more energy
d) Google Search servers use more energy

9. **What economic benefit does Microsoft's agreement bring to Pennsylvania?**
a) Reduced energy prices for residents
b) Job creation
c) Increased tourism revenue
d) Lowering state taxes

10. **What is a major challenge associated with AI server operations?**
a) Minimal data processing needs
b) High heat generation requiring extensive cooling
c) Lack of technological infrastructure
d) Limited computational power

Answers

1. a) TMI Energy Park
2. b) Access to nuclear energy
3. c) Nuclear
4. b) Achieve carbon neutrality
5. Answer: b) It has a high-capacity factor and provides baseload power
6. b) Solar
7. b) A nuclear incident

8. c) AI servers use significantly more energy
9. b) Job creation
10. b) High heat generation requiring extensive cooling

Are We Done Yet?

Almost

Thank you for joining me on this journey. I pulled content from six books in my *AI Ain't So Tough Series* at Amazon. Putting this together was my primary project in December, 2024. I even edited the *AI Disclosure* section on Christmas Day! I will publish updates every quarter.

I believe we are in a race. This race involves embracing and understanding AI. Those who do find themselves with a comparative advantage. They get more done in less time. The opposite is also true.

I also believe we will never stop learning. Just take it from a cybersecurity trainer, the more you know, the more you realize there is more to learn.

Mr. Sugartooth is a Persona

In case you're wondering, Mr. Sugartooth is an alternate persona. He is a super cool junior high school teacher who adds puns in otherwise serious content. Understanding AI is a tall order. Mr. Sugartooth helps make AI more consumable. After all, I named this series *AI Ain't So Tough*, not *AI is Too Tough*.

If you enjoyed reading this, please leave a positive review.

AI Disclosure

Some content here is generated by AI. The views expressed are my opinion and not legal advice. I am not a lawyer. You are an adult and responsible for anything you do.

You can find Mark Anthony Germanos on LinkedIn and Facebook.

Request for Review

Now that you've reached the end and savored the story, I have a small favor to ask. Would you consider sharing your thoughts at https://www.amazon.com/gp/product/B0DK41RWZS? Your review can make a world of difference to both the author and future readers. It only takes two minutes, but it means a lot. Simply head over to https://www.amazon.com/gp/product/B0DK41RWZS or point your cell phone camera at the QR Code (below), tap the link and let others know what you think. Whoa, did you notice the rhyme? Thank you. Thank you.

Your opinion matters. Your words could inspire someone else to discover their next favorite read. Thanks in advance.

You can also email me at mark@cybersafetynet.net, or point your phone's camera at this QR code (below) to contact me. I do not sell email addresses. I **will** alert you when I publish updates.

Your words could inspire someone else to discover their next favorite read. Thanks in advance. Do not be a stranger.

Bonus Content

How to tell someone is copying and pasting

More and more, we find people prompting AI for content and then mindlessly copying and pasting that AI-generated content online. It sticks out like a sore thumb. I used AI to generate parts of this book **but** I also spent a lot of time polishing that content before anyone else read it.

You can sometimes tell if content was written by AI without receiving any human polishing. It becomes obvious. Look for repetitive phrasing, slightly robotic sentence transitions and a lack of personal anecdotes. Unpolished AI writing often feels "safe." It is grammatically perfect but emotionally flat, like it was written by a computer. It reads a little too smooth to feel personal. You look up from the page and ask "Who wrote this blandness? A computer?"

A client sent me a document last week. The tone was far from his usual tone. I looked up from the page and asked "Who wrote this blandness? A computer? Surely not Rick."

In addition to the tone, here's what to look for:

- Em dashes. These pull sentences together. An example you'll see again is "The internet is our most powerful tool for connection and innovation—and it should stay open, fair, and free for all." Why is the "—" there anyway? I would have put a period after "innovation" and then started a new sentence with "It should stay open...."
- Since I'm bashing the em dash here, let me say AI often skips putting spaces in front of and behind em dashes. A person would write "The internet is our most powerful tool for connection and innovation — and it should stay open, fair, and free for all." AI would not put a space before or after the em dash.

That sentence would read as it does in the previous bullet point.

- Repetitive phrases: AI often repeats ideas or sentence structures.
- Odd transitions: The flow between sections can be abrupt or robotic.
- Generic language: Instead of personal anecdotes or vivid details, expect standard vocabulary and little emotion.
- Overly perfect grammar: Sometimes, the content lacks the little quirks or mistakes humans make. I intentionally avoid the Oxford comma. I am aware of the Maine court case, where the lack of an Oxford comma in a contract created ambiguity and a lawsuit. I still remove Oxford commas AI gives me.
- Missing context or perspective: AI text may skip nuanced opinions or local context a human expert would know.
- Too many emoticons. My English teachers did not teach emoticons. James Patterson does not use emoticons. AI does. Copying and pasting text with too many 🚀 ✨ 🌟 💼 💡 🌐 🛡️ 📊 🔒 💻 🛡️ 🩹 🥺 ❤️‍🩹 🐻 🧸 emoticons makes it look silly. Skip the emoticons—keep your professionalism. Did you see the dreaded em dash (above)? Let me try this again. Skip the emoticons. Keep your professionalism.

I am presenting AI-generated content and my human-polished content on three timely topics. I'm sure you'll notice the difference. I already presented these themes. Here, I am showing the difference between AI-generated content and content you polish.

Example 1: AI's 4-Day Workweek 🌟 💼

💡 Imagine reclaiming one more day every week just for YOU! The 4-day workweek boosts productivity, reduces burnout, and inspires creativity. Studies show employees are happier, businesses save on overhead, and teams work more effectively with focused schedules. More rest means more innovation. With three days for rest, passion projects, or family time, you don't just "live for the weekend" — you start living fully every day. Ready for happier Mondays and energized Fridays? Let's work smarter, not harder. The future of work is shorter, sharper, and way more satisfying!

Example 1: My 4-Day Workweek

Imagine having a three-day weekend every weekend. The 4-day workweek boosts productivity, reduces burnout and inspires creativity. People are happier, businesses save money and teams work more effectively with more focus. More rest means more reflection, passion projects and family time. Speaking of family time, did you know Japan's population is declining? Japan's government is pushing a four-day workweek so young people can spend more time at home and make babies. **I kid you not**! Work smarter, not harder.

Example 2: AI's Net Neutrality

Net neutrality means all internet traffic is treated equally—no "fast lanes" for those who pay more, no throttling your favorite streaming site, no blocking certain voices. Without it, big corporations could limit your access to ideas, products, and communities. With it, a student, entrepreneur, or artist can share their vision worldwide without unfair barriers. The internet is our most powerful tool for connection and innovation—and it should stay open, fair, and free for all. Let's protect digital democracy. Net neutrality isn't just policy; it's the foundation of an open web.

Example 2: My Net Neutrality

Net neutrality means all internet traffic gets equal treatment. It all travels at the same speed. Your ISP (Internet Service Provider), be it AT&T, Charter, Comcast, Google, Spectrum or whomever, cannot reduce speed of some content and more importantly, cannot filter content. All traffic gets equal treatment. With net neutrality, a student, entrepreneur or artist can share their vision worldwide without unfair barriers. The internet is our most powerful tool for connection and innovation. It should stay open, fair and unfiltered for everybody. Let's protect it.

Example 3: AI's Cyber Insurance 🔓 💻

🛡️ In today's digital world, one cyberattack can halt your business overnight. Phishing scams, ransomware, and data breaches are more common than ever—and recovery costs can be devastating. Cyber insurance gives you a safety net, covering financial losses, legal fees, and crisis management so you can bounce back faster. Think of it as health insurance for your data—you hope you'll never need it, but if disaster strikes, you'll be glad you have it. Your clients trust you with their information. Make sure you have the protection to keep that trust strong. 💪 📡

Example 3: My Cyber Insurance

Who am I kidding? I hate it. Cyber insurance is a bad place to put money. Invest time and money identifying your risks and reducing the likelihood that an attack would be successful. When your computer guy tells you to buy a firewall with a support subscription, do what he says.

Could you tell the difference?

Even if I did not add the headings, you could probably tell within 10 seconds which paragraphs I copied and pasted, and which ones I polished.

This is your challenge. Resist the urge to mindlessly copy and paste AI-generated text. Instead, polish what AI gives you. Make it read like you wrote it. If people read your work and respond, "It is bland. It reads like a computer wrote it," you've failed.

The Marketing Department

Since I'm a writer, most of my AI usage generates verbiage. That is not exclusive. I (and you can) use AI to generate other content.

"But you have a chapter on creating music. What else can we do?" Fasten your seatbelts. You can also use AI to generate visual content.

The original cover was rather busy. Same with my second cover. Although I was proud of both, they did not look good as a 1" thumbnail. I went to Perplexity and asked for help.

My prompt was "I am creating a book cover for a book called *AI for Beginners*. This will be sold at Amazon KDP. The final dimensions will be 6" wide x 9" high. Give me advice on a) the fonts and font sizes I should use, b) the colors I should use and c) what text I should add in addition to the title. Create three book cover mockups.

None were fantastic and none were awful. I did, however, realize that I liked bright blue or green on a dark cover.

I built a cover from scratch. Maybe the AI chip was the clincher. It is not original. I generated computer chip images at my preferred stock photo site, Freepik.

I prompted for a computer chip that says "AI." Focus on images that are blue. You know the image I ended up choosing for the cover. Some runners-up include:

Although strong, none of these wound up on the cover. Option 3 has a lowercase "i" and a capital "I" in place of the period. Not sure why that happened. You have to be careful with AI-generated content. AI mistakes, or hallucinations, are not limited to three-legged dancers.

The Amazon page

Maybe you read this book's entire Amazon page.[xxvi] Perplexity told me 64% of book buyers are women. I updated the A+ (visual content) to include graphics of women. One section has five women. The end result looks like this:

Maybe you're wondering where I found those pictures. Actually, I generated them at Freepik. My prompts included:

- Generate an image of a successful professional woman in a medical environment using AI. Give her a blue or green blouse.
- Give me a picture of a professional woman at a desk, using a computer.
- Give me a picture of a professional woman...

I'm sure you get the picture. Freepik proposed several. I merely picked and chose the ones I liked for the montage (above).

Let's talk about the leftmost picture...a woman doctor. The woman doctor was a stock image. I tweaked the image to place her in a hospital hallway. Freepik gave me several variations. Some made sense, some were good candidates and some just did not fit at all. I am biased against stock photos with the person looking at the camera. Several variations are online.[xxvii]

Remember, I started with stock pictures. All are strong, but all need new backgrounds. I prompted to add a hospital hallway background. The result with candidate #1 (top) in a hospital hallway is the AI-generated image I used.

I could have made her older or younger. I could have prompted for blond or red hair. I could have prompted for other demographics. What is important is that you have a library of stock images and Freepik (or whichever service you choose) lets you alter those images for your needs.

This is the Bonus section. No quiz here. Put this knowledge to good use. Check back often. AI is changing rapidly. Onward and upward.

Endnotes

[i] https://www.dictionary.com/browse/artificial%20intelligence
[ii] Sometimes I refer to Perplexity, sometimes I refer to Perplexity.ai and sometimes I refer to https://perplexity.ai. Same service.
[iii] https://en.wikipedia.org/wiki/Flesch%E2%80%93Kincaid_readability_tests#Flesch%E2%80%93Kincaid_grade_level
[iv] https://www.goldengate.org/assets/1/6/retro3.pdf.
[v] https://www.forbes.com/sites/brittanyanas/2024/08/31/the-20-best-us-tourist-attractions-ranked-according-to-a-2024-report/.
[vi] https://surefirecpr.com/cpr/life-saving-beats-the-ultimate-cpr-song-playlist-for-emergency-situations/ and https://www.procpr.org/blog/training/cpr-chest-compression-rate
[vii] https://www.youtube.com/watch?v=qYkbTyHXwbs
[viii] https://suno.com/song/cac22c27-fd70-4ec8-a5fc-92c7003ee78c
[ix] https://suno.com/song/365ee095-78bf-4068-8a15-bf45e69a9d93
[x] https://suno.com/song/a0e3d033-1aa3-4ab4-9d72-1a98439accb6
[xi] https://suno.com/song/1532f546-36c8-4acc-b433-71e2520df0fc
[xii] https://suno.com/song/5563e8e0-8728-465e-a859-2a5030d87bb6
[xiii] https://suno.com/song/5413eebf-f5cd-4f71-b640-279a21d5a653
[xiv] https://suno.com/playlist/d6f73009-5c71-48b3-85a4-e00b84f7e472.
[xv] https://www.euronews.com/culture/2025/03/18/made-in-spain-the-first-music-album-created-entirely-with-artificial-intelligence
[xvi] https://www.analyticsvidhya.com/blog/2024/04/deepfake-videos-that-you-must-watch/
[xvii] https://www.theguardian.com/us-news/2024/jan/22/biden-fake-robocalls-new-hampshire
[xviii] https://incode.com/blog/top-5-cases-of-ai-deepfake-fraud-from-2024-exposed/
[xix] https://www.hollywoodreporter.com/business/business-news/writers-strike-ai-chatgpt-1235478681/
[xx] https://www.google.com/search?q=A+Recent+Entrance+to+Paradise
[xxi] https://www.perplexity.ai/search/tell-me-why-my-company-should-iPYEOwCzQ5uasPrKAzu3CA

[xxii] https://www.perplexity.ai/search/what-is-net-neutrality-oOYc4F7sSFapRYqleiMLYw

[xxiii] https://www.perplexity.ai/search/what-is-cyberinsurance-70HVmLimRZ6wKUHilSIljA

[xxiv] https://informaconnect.com/academy/companies-entering-trillion-dollar-club-in-2024/

[xxv] https://mepacademy.com/data-center-hvac-systems/

[xxvi] https://www.amazon.com/dp/B0DK41RWZS.

[xxvii] https://www.freepik.com/premium-psd/freepik_403355756.htm

Made in the USA
Monee, IL
04 November 2026

67383077R00118